Field Sampling Methods

for

Remedial Investigations

Mark E. Byrnes

Contributors
Donna M. Leydorf
David B. Smet

LEWIS PUBLISHERS
Boca Raton   Ann Arbor   London   Tokyo

Library of Congress Cataloging-in-Publication Data

Byrnes, Mark E.
    Field sampling methods for remedial investigations / Mark E. Byrnes ; contributors,
Donna M. Leydorf, David B. Smet.
      p. cm
    Includes bibliographical references and index.
    1. Soil pollution — Measurement — Technique.   2. Water — Pollution — Measurement —
Technique.   3. Soils — Sampling — Technique.   4. Water — Sampling — Technique.   5.
Groundwater — Sampling — Technique.
    I. Leydorf, Donna M. II. Smet, David B. III. Title.
TD878.B96   1994
628.5′5′0287 — dc20   93–40297
ISBN 0-87371-698-1

This book contains information obtained from authentic and highly regarded sources. Reprinted material is quoted with permission, and sources are indicated. A wide variety of references are listed. Reasonable efforts have been made to publish reliable data and information, but the author and the publisher cannot assume responsibility for the validity of all materials or for the consequences of their use.

No claim to original U.S. Government works
International Standard Book Number 0-87371-698-1
Library of Congress Card Number 93-40297
Printed in the United States of America   3 4 5 6 7 8 9 0
Printed on acid-free paper

Preface

This book seeks to provide the reader with guidance on the development of an effective field sampling program, and Standard Operating Procedures (SOPs) and/or technical information and guidance for many of the most effective remedial investigation methods. While this book focuses on intrusive investigative techniques, nonintrusive techniques such as aerial photography, surface geophysics, and surface radiological surveying have also been addressed, but in less detail.

SOPs have been provided for those sampling techniques which do not require specialized academic training, such as soil, sediment, surface water, groundwater, and drum sampling. For more specialized investigative techniques such as underground drainage surveying and some types of soil-gas surveying, information has been provided to help the reader to understand how the technique works and under what conditions it can be used most effectively. This book focuses on those methods and procedures which have both proved themselves to be effective, and are acknowledged by the U.S. Environmental Protection Agency (EPA) as reputable techniques.

The primary EPA guidance documents which have been used to support the procedures and technologies recommended in this book include:

- A Compendium of Superfund Field Operations Methods, EPA/540/P-87/001a, 1987
- Characterizing Heterogeneous Wastes, EPA/600/R-92/033, 1992
- Data Quality Objectives for Remedial Response Activities, EPA/540/G-87/003, 1987
- Draft Standard Operation Procedures for Field Samplers, Region VIII, Revision 4, 1992
- Drum Handling Practices at Hazardous Waste Sites, EPA/600/S2-86/013, 1986
- Environmental Compliance Branch Standard Operating Procedures and Quality Assurance Manual, Region IV, 1991
- Final Comprehensive State Ground Water Protection Program Guidance, 100-R-93-001, 1992
- Guidance for Conducting Remedial Investigations and Feasibility Studies Under CERCLA, EPA/540/G-89/004, 1988
- Guidance on Remedial Investigations Under CERCLA, EPA/540/G-85/002, 1985
- Guide to Management of Investigation-Derived Wastes, PB92-963353, 1992

- Handbook of Sampling and Sample Preservation of Water and Wastewater, EPA/600/4-89/034, 1982
- Handbook of Suggested Practices for the Design and Installation of Ground-Water Monitoring Wells, EPA/600/4-89/034, 1991
- Management of Investigation-Derived Waste During Site Inspections, EPA/540/G-91-009, 1991
- RCRA Comprehensive Ground-Water Monitoring Evaluation Document, OSWER-9950.2, 1988
- RCRA Facility Investigation (RFI) Guidance, Volumes I, II, III, and IV, PB89-200299, 1989
- RCRA Ground-Water Monitoring: Draft Technical Guidance, PB93-139350, 1992
- RCRA Ground-Water Monitoring Technical Enforcement Guidance Document, OSWER-9950.1, 1986
- Soil Sampling Quality Assurance User's Guide, EPA/600/8-89/046, 1989
- USEPA Contract Laboratory Program Statement of Work for Organic/Inorganic Analysis, OLM01.0/ILM02.1, 1990/1991

This book has been written for field geologists, hydrogeologists, and other environmental scientists who are involved in the development and implementation of remedial investigation studies. Others who can benefit from this book include environmental regulators who need a book that provides procedures that are consistent with available EPA guidance, and educators and students who are interested in the practical elements of developing and implementing an environmental sampling program.

The author would appreciate criticisms or suggestions, should anyone note errors or have suggestions on how later editions may be improved. Although he has attempted to provide detailed coverage of sampling principles and methods, some emerging technologies may have been overlooked.

Mark E. Byrnes is a Senior Environmental Geologist, working under Science Application International Corporation's (SAIC's) Environmental Compliance Group in Oak Ridge, Tennessee. He has experience leading remedial investigation programs and writing Field Sampling Plans and Remedial Investigation Reports for the U.S. military, Department of Energy (DOE), and private industry installations across the country, as well as in Italy and Guam. Mr. Byrnes received a Bachelor of Arts degree in Geology from the University of Colorado, Boulder, and a Master of Science degree in Geology from Portland State University, Portland, Oregon. He is currently serving as SAIC's Remedial Investigation Specialist under DOE's Formerly Utilized Sites Remedial Action Program (FUSRAP).

Science Applications International Corporation is a 12,000-person multidisciplined corporation with 20 years of experience supporting the federal government on important environmental programs. Approximately 25% of SAIC's business is in the environmental arena. SAIC supports the Environmental Protection Agency by assisting in the operation of the Resource Conservation and Recovery Act (RCRA) Hotline and by performing technical reviews on environmental reports. SAIC also has large contracts with the Department of Energy (DOE) and Martin Marietta Energy Systems, Inc., performing environmental studies at DOE facilities and military installations across the United States.

Contributors

Donna M. Leydorf is a Senior Attorney working in SAIC's Environmental Analysis Division in Oak Ridge, Tennessee. She has been admitted to practice in South Carolina and Ohio, and her admission to Tennessee is pending. Ms. Leydorf is an experienced litigator and has expertise in land use as well as environmental issues. She received her Bachelor of Arts degree in psychology from Midwestern State University, Texas, and her Juris Doctorate degree from the University of South Carolina. She currently provides compliance support and legal analysis for remediation projects.

David B. Smet manages the Engineering Surveillance and Systems Group for Westinghouse Hanford Company in Richland, Washington. His background and education are in biomedical and communications electronics, with a specialty in remote ultrasonic, eddy current, and visual systems. Mr. Smet is currently providing remote characterization systems and support to the Hanford Site for environmental and waste management programs.

List of Acronyms and Abbreviations

AOC	area of contamination
ARARs	Applicable Relevant and Appropriate Requirements
ASTM	American Society of Testing and Materials
BGS	below ground surface
C°	degrees Centigrade
CERCLA	Comprehensive Environmental Response, Compensation, and Liability Act
CAA	Clean Air Act
CCD	charged coupled device
CFR	Code of Federal Regulations
CGI	combustible gas indicator
CID	Charge Injection Device
CLP	Contract Laboratory Program
COC	chain-of-custody
cpm	counts-per-minute
CPT	cone penetrometer
CRT	cathode ray tube
CWA	Clean Water Act
CX	Categorical Exclusion
DNAPLs	dense nonaqueous phase liquids
DOT	U.S. Department Of Transportation
DOE	U.S. Department Of Energy
DPM	Direct Push Method
DSE	Domestic Sewage Exclusion
DQOs	Data Quality Objectives
EA	Environmental Assessment
EIS	Environmental Impact Statement
ELR	Environmental Law Reporter
EPA	U.S. Environmental Protection Agency
ft	feet
F°	degrees Fahrenheit
FID	flame ionization detector
FONSI	Finding Of No Significant Impact
FR	Federal Register
FS	Feasibility Study

FSP	Field Sampling Plan
FUSRAP	Formerly Utilized Sites Remedial Action Program
gpm	gallons-per-minute
HIGI	Hanford Information Gathering Instrument
hr	hour
H&S	Health and Safety
IDLH	Immediately Dangerous to Life and Health
IDW	Investigation Derived Waste
lb	pound(s)
LDRs	Land Disposal Restrictions
LNAPLs	light non-aqueous phase liquids
MCL	Maximum Contaminant Level
MCLG	Maximum Contaminant Level Goal
min	minute
mL	milliliter
mL/min	milliliter-per-minute
NCP	National Contingency Plan
NEPA	National Environmental Policy Act
NOI	Notice Of Intent
NPDWRs	National Primary Drinking Water Regulations
NPL	National Priorities List
NTU	Nephelometric Turbidity Units
O.D.	outside diameter
ONPA	Office of NEPA Project Assistance
OSHA	Occupational Safety and Health Administration
PARCC	precision, accuracy, representativeness, completeness, and comparability
PA/SI	Preliminary Assessment/Site Investigation
PCBs	polychlorinated biphenyls
pCi/gm	picocuries-per-gram
PID	Photoionization Detector
POTW	Publicly Owned Treatment Works
ppb	parts-per-billion
PPE	personal protective equipment
ppm	part-per-million
PRG	Preliminary Remediation Goal
psi	pounds-per-square-inch
PVC	polyvinyl chloride
QAPP	Quality Assurance Project Plan
QA/QC	Quality Assurance/Quality Control
RA	Remedial Action
RAD	Roentgen Absorbed Dose
RD	Remedial Design
RI	Remedial Investigation

RI/FS	Remedial Investigation/Feasibility Study
RCRA	Resource Conservation and Recovery Act
RFI	RCRA Facility Investigation
RPM	Regional Project Manager
SAIC	Science Applications International Corporation
SAP	Sampling and Analysis Plan
SARA	Superfund Amendments and Reauthorization Act
SCBA	Self Contained Breathing Apparatus
SDWA	Safe Drinking Water Act
SOPs	Standard Operating Procedures
TCLP	Toxic Characteristic Leaching Procedure
TPH	total petroleum hydrocarbons
TSCA	Toxic Substances Control Act
TSD	treatment, storage, and disposal
UEL	upper explosive limit
U.S.	United States
USGS	U.S. Geological Survey
UST	Underground Storage Tank
USC	U.S. Code
VOA	volatile organic analysis
VOC	volatile organic compound

Acknowledgments

The author would like to acknowledge SAIC's Environmental Compliance Group, led by Dr. Barry Goss, which provided funding to support the preparation of tables and figures for this book. Specific acknowledgment goes to Matthew Shafer, Doug Combs, Kevin Newman, Alfred Wickline, Duncan Moss, William Kegley, Patty Stoll, and Jeff Slack with SAIC, who provided technical reviews on various portions of the document. Other SAIC employees who provided valuable input included Brian Damiata, Steve Lanter, and Jerry Truitt. Technical input from Susan Spencer (Environmental Instruments), Randy Rohrman (EPA Region VII), Jon Novick (Bechtel National, Inc.), Ken Skinner (Bechtel National, Inc.), Steve Kautz (Bechtel National, Inc.), and Robert Pirkle (Microseeps) is also greatly appreciated. The tables and figures were generated with the assistance of Stacy Heptinstall and Denise Haney. Finally, the author would like to express his appreciation to his wife Karen, daughters Christine and Kathleen, and his mother and father for their support throughout the preparation of this document.

Contents

Field Sampling Methods

for

Remedial Investigations

CHAPTER 1

Introduction

This book has been written with the objective of providing guidance on how to develop an effective field sampling program, and provides Standard Operating Procedures (SOPs) and/or technical guidance for many of the most effective remedial investigation tools available on the market today. The guidelines provided are slanted toward the Comprehensive Environmental Response, Compensation, and Liability Act (CERCLA) Remedial Investigation/Feasibility Study (RI/FS) program, but are also useful for the Resource Conservation and Recovery Act (RCRA) Facility Investigation (RFI) program. This book focuses on those methods and procedures which have both proved themselves to be effective, and are accepted as reputable techniques by the U.S. Environmental Protection Agency (EPA).

To assist in the development of a field sampling program, Chapter 2 outlines the three stages of the EPA's Data Quality Objectives (DQOs) process, which involves identifying decision types (Stage 1), identifying decision uses and needs (Stage 2), and designing a data collection program (Stage 3). Under that chapter, guidance is provided to assist in:

- developing a site conceptual model
- outlining sampling objectives
- identifying data quality and quantity needs, and
- designing and developing a data collection program.

To assist in selecting the most appropriate site-specific sampling techniques, summary tables have been provided which rate various techniques against their effectiveness in collecting samples for specific laboratory analyses, sample types, and sampling depths. A discussion is also provided on various types of supplemental characterization tools which are commonly used to assist in defining the aerial extent of contamination.

Specific details on recommended technologies and sampling tools are provided in Chapter 3, along with information regarding their effectiveness and limitations. SOPs have been provided in that chapter for those sampling techniques which do not require specialized academic training, such as soil, sediment, surface water, and groundwater sampling. For more specialized investigative techniques such as surface geophysical surveying, and underground drainage surveying, information has been provided to assist the reader in understanding how the technique works, and under which conditions it can be used most effectively.

Chapters 4 through 7 provide additional information which is critical to the effectiveness of a field sampling program, including: equipment decontamination; sample preparation, documentation, and shipment; health and safety; and management of investigation-derived waste. All of these chapters have been written based on practical field experience, along with EPA and other regulatory guidance.

To add to the usefulness of this book as a field or office reference manual, a short summary of many of the United States, primary environmental laws has been provided, along with general reference tables, and a bibliography at the end of several chapters to provide the reader with supplemental references.

With the information provided, one should be able to develop thorough and cost-effective field sampling programs using some of the most current and effective field sampling tools. Since the recommended sampling methodologies have been developed with the assistance of many of the most recent EPA guidance manuals, they are recommended for all sites, particularly those on the National Priorities List (NPL).

GENERAL CONSIDERATIONS

Prior to beginning the development of a field sampling program, a project manager should pull together his/her key staff members representing all the various disciplines required for the remedial process. These disciplines include, but are not limited to, hydrogeology, biology, toxicology, health physics, and geotechnical and environmental engineering. As a team, these members should evaluate all of the available data and background information about the site. Developing a field sampling program with this expertise will assure that all of the key data needs are addressed. Once this evaluation has been completed, the following questions should be asked:

- What environmental media sampling is required?
- What flora/fauna sampling is required?
- How many samples need to be collected?
- What sampling schemes are most appropriate?
- What sampling methods are most appropriate?
- What analytical methods are most appropriate?
- What are the data quality requirements? and
- What supplemental characterization tools can be used to reduce cost?

Each of the above questions are addressed as part of the DQO process outlined in Chapter 2.

In addition to the technical requirements of a field sampling program, one must also consider the regulatory aspects. For example, regulatory requirements on how to properly handle and dispose of investigation-derived waste, and how to properly ship environmental samples, must be met. Other regulatory considerations include meeting health and safety training and medical surveillance requirements. These requirements are addressed in Chapters 5, 6, and 7. The following section also addresses some of these issues, but in more general terms.

SUMMARY OF MAJOR ENVIRONMENTAL LAWS

Soon after Congress passed the Comprehensive Environmental Response, Compensation, and Liability Act (CERCLA) in 1980, which was later modified by the Superfund Amendments and Reauthorization Act (SARA) in 1986, environmental investigations began at abandoned or closed chemical and radiological waste sites across the country. Currently, billions of dollars are being spent each year on the characterization and remediation of these and other sites.

Other environmental laws which are currently working along with CERCLA/SARA to regulate environmental site operations, industrial activities, and emissions include the Resource Conservation and Recovery Act (RCRA), Toxic Substance Control Act (TSCA), National Environmental Policy Act (NEPA), Clean Water Act (CWA), Safe Drinking Water Act (SDWA), and Clean Air Act (CAA). While each of these laws is different in terms of the types of materials or activities that they regulate, together they are our nation's means of controlling the quality of our environment (Figure 1.1).

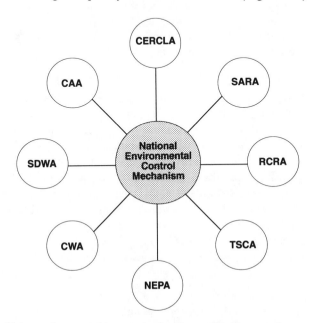

Figure 1.1.   Major environmental laws protecting the quality of our environment.

Due to the complexity and ever-changing requirements of environmental laws, most companies hire environmental lawyers to assist them with compliance issues. The following information has been provided to the reader for general guidance on the types of issues addressed under each of the above-mentioned laws, and was derived primarily from a Statute Binder published by the Environmental Law Reporter (1990), and other regulatory guidance manuals (see References). This information is not meant to be cited as law.

CERCLA Compliance

In 1980, Congress passed CERCLA (42 USC §§9601–9675), which regulates the cleanup of abandoned or closed waste sites across the country, in addition to providing requirements and guidance to respond to unpermitted and uncontrolled releases of hazardous substances into the environment. CERCLA utilizes the RI/FS process to evaluate the environmental conditions at a site, and to select the final remedial alternative which will be implemented to remediate the site (EPA 1988, Figure 1.2).

The RI/FS process consists of scoping, site characterization, development and screening of remedial alternatives, treatability investigations, and detailed evaluation of remedial alternatives. Scoping is the initial planning phase of site remediation that selects a general approach for managing the site. Specific project plans are developed to:

- determine the types of decisions to be made
- identify the data quality objectives needed to support these decisions
- describe the methods by which data will be obtained and analyzed, and
- prepare project plans to document methodology.

Site characterization involves conducting field investigations, performing laboratory analyses on environmental samples, and evaluating the results for the purpose of identifying the types, concentrations, sources, and extent of contamination at a site. These data are then used in the development of a Baseline Risk Assessment, and to evaluate potential remedial alternatives. Site characterization studies commonly use tools such as soil-gas surveying, and surface and downhole geophysics, in combination with the sampling of environmental media. For cost-effectiveness, field sampling programs are commonly phased, whereas secondary sampling phases are used to better define the results obtained in the primary phase (EPA 1988).

Preliminary remedial alternatives should be developed when likely response scenarios are first identified. The development of alternatives requires the:

- identification of remedial action objectives
- identification of potential treatment, resource recovery, containment, and disposal technologies that will satisfy these objectives

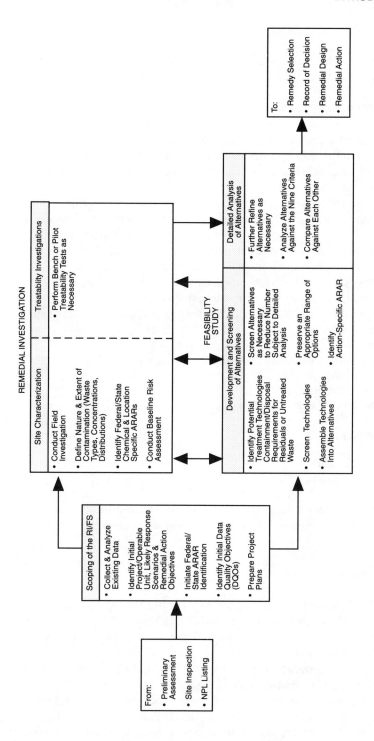

Figure 1.2. Outline of the CERCLA RI/FS process.

- screening of technologies based on their effectiveness, implementability, and cost, and
- assembling of technologies and their associated containment or disposal requirements into alternatives for each contaminated media.

A range of treatment alternatives should be considered. The range should vary primarily in the extent to which the alternatives rely on long-term management. The upper bound of the range should be an alternative that would eliminate the need for any long-term management, while the lower bound should consist of an alternative that involves little or no treatment, such as no-action, or long-term monitoring.

Treatability studies provide sufficient data to develop and evaluate treatment alternatives to support the remedial design of the selected alternatives. They also reduce cost and performance uncertainties to help select a remedy.

If no treatability data are available for a site, treatability tests may be necessary to evaluate the effectiveness of a particular technology to that site. Treatability tests typically involve bench-scale testing, followed by pilot-scale testing if a technology appears to be feasible (EPA 1988).

Once sufficient data are available, alternatives are evaluated in detail against the following nine evaluation criteria (EPA 1988):

- overall protectiveness to human health and the environment
- compliance with ARARs
- long-term effectiveness
- reduction of toxicity, mobility, or volume
- short-term effectiveness
- implementability
- cost
- state acceptance, and
- community acceptance.

Each alternative is analyzed against each of the above criterion and then compared against one another to determine their respective strengths and weaknesses.

SARA Compliance

In 1986, CERCLA (42 USC §§9601–9675) was modified by the Superfund Amendments and Reauthorization Act (SARA). SARA did not change the basic structure of CERCLA, but rather modified many of the existing requirements, and added a few new ones.

SARA places a strong statutory preference on remedial alternatives that are highly reliable and provide long-term protectiveness. A preference is placed on alternatives that employ treatment which permanently and significantly reduces the volume, toxicity, or mobility of contaminants. Offsite transport and disposal of contaminated materials is not a favored alternative when practicable treatment or resource recovery technologies are available. After the initiation of an action, SARA requires a review of the remediation's effectiveness at least every 5 years, for as long as contaminants may pose a threat to human health or the environment (ELR 1990).

Before the passing of SARA, no laws specifically protected the health and safety of hazardous waste workers; consequently, workers were at the mercy of their employers' safety policies. SARA addresses the risk of exposure to hazardous wastes and the need to protect employees exposed to these materials under 29 CFR 1910.120. Workers who are covered by this regulation include those involved with cleanup, treatment, storage, and disposal, and emergency response.

To ensure the protection of workers in hazardous environments, a Health and Safety Program must be implemented. As part of this program, a Health and Safety Plan must be written to identify, evaluate, and control health and safety hazards and provide for emergency response. The responsibility for developing this program and writing this plan rests on the employer of the hazardous waste workers.

As part of a health and safety program, medical surveillance is mandatory for all workers who (ELR 1990):

- are exposed to hazardous substances or health hazards at or above the permissible exposure limits,
- wear a respirator for 30 days or more a year,
- are injured, become ill, or develop signs or symptoms due to possible overexposure involving hazardous substances from an emergency response or hazardous waste operation, or
- belong to a Hazardous Material team.

SARA requires employers to establish a program to inform all personnel involved in a hazardous waste operation of the nature, level, and degree of exposure that can be anticipated as a result of their participation. Other key issues which SARA addresses include: the handling of drums and containers; procedures for decontamination; emergency response; illumination and sanitation.

RCRA Compliance

In 1976, Congress passed the Resource Conservation and Recovery Act (RCRA), which regulates Solid and Hazardous Waste, prevents new uncontrolled hazardous sites from developing, and protects human health and the environment. The RCRA requires the "cradle to grave" management of hazardous waste.

Under RCRA, the term "Waste" refers to any discarded material which is either abandoned, disposed of, burned or incinerated, or stored in lieu of being abandoned. Solid Waste materials include:

- garbage
- refuse
- sludge from a waste treatment plant, water supply treatment plant, or air pollution control facility, and
- other discarded material.

Other discarded material may include solid, liquid, semisolid, and contained gaseous material resulting from industrial, commercial, mining, agricultural operations, and from community activities. This material does not include solid or dissolved material in domestic sewage, solid or dissolved materials in irrigation return flows, industrial discharges, special nuclear, or by-product material (ELR 1990).

Hazardous Wastes are Solid Wastes which because of quantity, concentration, or physical, chemical, or infectious characteristics, may:

- cause or contribute to increase in mortality or increase in serious irreversible, or incapacitating reversible illness,
- pose substantial present or potential hazard to human health or environment if improperly treated, stored, transported, or disposed of.

A waste can be defined as "Hazardous" if it meets one of the following criteria:

- Ignitable: flash point of 140° F or less
- Corrosive: pH of 2 or less (acid); 12.5 or greater (base)
- Reactive: unstable, capable of detonation, explosive
- Toxic: as determined by Toxicity Characteristic Leaching Procedure (TCLP).

Table 1.1 presents the regulatory levels for each of the various contaminants tested under TCLP. If the TCLP results for a soil exceed any of these levels, the soil is defined as a RCRA Hazardous Waste. The following materials are not considered Hazardous Waste (ELR 1990):

- household, farming, mining, and fly ash wastes
- drilling fluids used in oil and gas exploration
- waste failing TCLP for chromium if several tests were run and only one sample failed
- solid waste from processing ores
- cement kiln dust
- discarded wood products failing TCLP for arsenic
- petroleum contaminated media failing TCLP if remediation under UST rules.

RCRA also regulates medical waste, tanks, tank systems, surface impoundments, waste piles, land treatment facilities, landfills, incinerators, generators, and other types of miscellaneous units such as boilers, industrial furnaces, and underground injection wells.

Table 1.1 Maximum Concentration of Contaminants for the Toxicity Characteristic

EPA HW No.[a]	Contaminant	CAS No.[b]	Regulatory Level (mg/L)
D004	Arsenic	7440–38–2	5.0
D005	Barium	7440–39–3	100.0
D018	Benzene	71–43–2	0.5
D006	Cadmium	7440–43–9	1.0
D019	Carbon tetrachloride	56–23–5	0.5
D020	Chlordane	57–74–9	0.03
D021	Chlorobenzene	108–90–7	100.0
D022	Chloroform	67–66–3	6.0
D007	Chromium	7440–47–3	5.0
D023	o-Cresol	95–48–7	200.0[d]
D024	m-Cresol	108–39–4	200.0[d]
D025	p-Cresol	106–44–5	200.0[d]
D026	Cresol		200.0[d]
D016	2,4-D	94–75–7	10.0
D027	1,4-Dichlorobenzene	106–46–7	7.5
D028	1,2-Dichloroethane	107–06–2	0.5
D029	1,1-Dichloroethylene	75–35–4	0.7
D030	2,4-Dinitrotoluene	121–14–2	0.13[c]
D012	Endrin	72–20–8	0.02
D031	Heptachlor (and its epoxide)	76–44–8	0.008
D032	Hexachlorobenzene	118–74–1	0.13[c]
D033	Hexachlorobutadiene	87–68–3	0.5
D034	Hexachloroethane	67–72–1	3.0
D008	Lead	7439–92–1	5.0
D013	Lindane	58–89–9	0.4
D009	Mercury	7439–97–6	0.2
D014	Methoxychlor	72–43–5	10.0
D035	Methyl ethylketone	78–93–3	200.0
D036	Nitrobenzene	98–95–3	2.0
D037	Pentachlorophenol	87–86–5	100.0
D038	Pyridine	110–86–1	5.0[c]
D010	Selenium	7782–49–2	1.0
D011	Silver	7440–22–4	5.0
D039	Tetrachloroethylene	127–18–4	0.7
D015	Toxaphene	8001–35–2	0.5
D040	Trichloroethylene	79–01–6	0.5
D041	2,4,5-Trichlorophenol	95–95–4	400.0
D042	2,4,6-Trichlorophenol	88–06–2	2.0
D017	2,4,5-TP (Silvex)	93–72–1	1.0
D043	Vinyl chloride	75–01–4	0.2

[a]Hazardous waste number.
[b]Chemical abstracts service number.
[c]Quantitation limit is greater than the calculated regulatory level. The quantitation limit therefore becomes the regulatory level.
[d]If o-, m-, and p-cresol concentrations cannot be differentiated, the total cresol (D026) concentration is used. The regulatory level of total cresol is 200 mg/L. (source: 40 CFR Part 261, Appendix II).

Some of the waste generators' responsibilities under RCRA include knowing what types of waste materials are being generated at a particular site through either knowledge or testing. The generator must obtain an EPA identification number after notifying them of the type and status of the waste being generated. All Hazardous Waste must be consolidated at a permitted Hazardous Waste storage unit, or a satellite accumulation point, where this waste can be temporarily stored for 90 days before transport to a treatment storage and disposal facility (ELR 1990).

Prior to transporting the Hazardous Waste material, the EPA must be notified of the proposed activity, proper manifesting requirements must be met, and emergency response procedures must be in place. A copy of the manifest must be kept on file to document that waste was properly disposed of at a permitted facility. Any transportation of Hazardous Waste must comply with all Department of Transportation regulations.

All treatment, storage, and disposal (TSD) facilities must use Part A permit applications, and must comply with general facility standards, interim status technical standards, closure/post-closure standards and notification requirements. The Part A permit application identifies the type of Hazardous Waste managed, estimates the annual quantities of material managed, provides details on methods of waste management, and includes a facility map.

Each TSD facility is required to develop and implement a waste analysis plan to test incoming Hazardous Waste, which serves to ensure that the waste material received matches the manifest. They must also provide appropriate security measures, conduct regular facility inspections, implement a groundwater monitoring program, provide appropriate personnel training, and keep operating records. The operating records must include information such as a description and quantity of waste received, results of waste analysis testing, inspection findings, summary reports of incidents, closure and post-closure cost estimates, and land disposal restriction certifications, and notifications.

RCRA requires that closure of a Hazardous Waste unit must begin within 90 days of the last receipt of Hazardous Waste, or upon approval of the closure plan, whichever is later. Post-closure care of a unit is required if closure in place is used (ELR 1990).

TSCA Compliance

The Toxic Substance Control Act (TSCA, 15 USC §§2601–2671) requires all manufacturers, processors, and distributors to maintain records of the hazards that each of their products pose to human health and the environment. It also requires the EPA to compile and publish a list of each chemical substance manufactured or processed in the United States. The statute authorizes the EPA to conduct limited inspections of areas where substances are processed, or stored, and of conveyances used to transport the substances.

TSCA requires all manufacturers and processors of new substances, or substances that will be applied to a significant new use, to notify the Administrator of the EPA that they intend to manufacture or process the substance. If analytical testing is required, the manufacturer or processor must provide the results along with the notification.

If the EPA finds the analytical testing to be insufficient, a proposed order is written to restrict the manufacturing of the substance until adequate testing is completed. If the testing data indicate that the substance may present a significant risk of cancer, gene mutations, or birth defects, the EPA will promulgate regulations concerning the distribution, handling, and labeling of the substance. In the case of an imminently hazardous substance, the EPA may commence a civil action for seizure of the substance, and possibly a recall and repurchase of the substance previously sold (ELR 1990).

The requirements of this statute generally do not apply to toxic substances distributed for export unless they would cause an unreasonable risk of harm within the United States. On the other hand, imported substances are subject to the requirements of the statute, and any substances that do not comply will be refused entry into the United States. Violations of this statute can result in both civil and criminal penalties, and the violating substance may be seized.

Some important regulations under TSCA govern the manufacture, use and disposal of polychlorinated biphenyls (PCBs). PCBs are found in many substances, such as oils, paints and contaminated solvents. The regulations establish concentration limits and define acceptable methods of disposal. PCBs may now be used only in totally enclosed systems.

TSCA also requires that asbestos inspections be performed in school buildings to define the appropriate level of response actions. The statute also requires the implementation of maintenance and repair programs, and periodic surveillance of school buildings where asbestos is located, as well as prescribing standards for the transportation and disposal of this material. For those school buildings containing asbestos, local educational agencies are required to develop an asbestos management program which must include plans for response actions, long-term surveillance, and use of warning labels for asbestos remaining in the buildings.

In an attempt to control radon contamination inside buildings, the EPA is required by this statute to publish a document titled "A Citizen's Guide to Radon," which includes information on the health risks associated with exposure to radon, the cost and technical feasibility of reducing radon concentrations, the relationship between long-term and short-term testing techniques, and outdoor radon levels around the country. This statute also requires the EPA to determine the extent of radon contamination in the nation's schools, develop model construction standards and techniques for controlling radon levels within new buildings, and make grants available to states to assist them in the development and implementation of their radon programs (ELR 1990).

NEPA Compliance

The National Environmental Policy Act (NEPA, 42 USC §§4321–4370a) was passed in 1969, and was one of the first statutes directed specifically at protecting the environment. NEPA documentation is necessary when any "major federal action" that may have a significant impact on the environment may be undertaken. The NEPA process places heavy emphasis on public involvement. Public notice must be provided for NEPA-related hearings, public meetings, and to announce the availability of environmental documents. In the case of a NEPA action of national concern, notice is included in the *Federal Register* and notice is made by mail to national organizations reasonably expected to be interested in the matter (ONPA 1988).

The primary documents prepared under the NEPA process are the Notice Of Intent (NOI), Environmental Impact Statement (EIS), Environmental Assessment (EA), Finding Of No Significant Impact (FONSI), and Categorical Exclusion (CX). Any environmental document in compliance with NEPA may be combined or integrated with any other agency document to reduce duplication and paperwork.

Before preparing an EIS, a NOI must be issued for public review. The NOI describes the proposed action and possible alternatives, describes the federal agency's proposed scoping process including whether, when, and where any public scoping meetings will be held, and finally states the name and address of a person within the agency who can answer questions about the proposed action.

The EIS serves as an action-forcing device to ensure that the policies and goals defined in NEPA are infused into the ongoing programs and actions of the federal government. The objective of the EIS is to provide a full and fair discussion of significant environmental impacts, and is used to inform decision-makers and the public of the reasonable alternatives which would avoid or minimize adverse impacts or enhance the quality of the human environment. The EIS is meant to serve as the means of assessing the environmental impact of proposed federal agency actions, but is not used to justify decisions which have already been made (ONPA 1988).

The EIS and other NEPA documents should be written so the public can readily understand them. Wherever there is incomplete or unavailable information, it is critical to overtly state this in the document. No decision on the proposed action shall be made or recorded under a federal agency until the later of the following dates: 90 days after publication of the notice for a draft EIS, or 30 days after publication of the notice for a final EIS (ONPA 1988).

An EA is a concise public document that determines whether to prepare an EIS. If there are no significant impacts on the environment, a FONSI is published. An EA can facilitate preparation of an EIS when one is needed, but is not necessary if it is already known that there will be significant impacts, and an EIS must be prepared.

A FONSI is a document prepared by a federal agency to briefly describe the reasons why an action will not have a significant effect on the human environment, and for which an EIS is not needed. This document includes the EA or a summary of this study, and notes any other environmental documents related to it. If the EA is included, the finding need not repeat any of the discussion in the assessment, but may incorporate it by reference.

A CX refers to a category of actions which do not individually or cumulatively have a significant effect on the human environment and which have been found to have no such effect in procedures adopted by a federal agency in implementation of these regulations. Consequently, there is no need for the preparation of an EA or an EIS.

Where emergency circumstances make it necessary to take an action with significant environmental impact without observing the provisions of these regulations, the federal agency taking the action should consult with the Council about alternative arrangements. Agencies and the Council will limit such arrangements to actions necessary to control the immediate impacts of the emergency. Other actions remain subject to NEPA review (ONPA 1988).

CWA Compliance

The objective of the Clean Water Act (CWA, 33 USC §§1251–1387) is to restore and maintain the chemical, physical, and biological integrity of the nation's waters. This act established interim water quality goals which provide for the protection of human health and the environment, and assure the propagation of fish, shellfish, and wildlife until the pollutants can be completely eliminated. The act seeks to eliminate the discharge of pollutants into navigable waters, in addition to addressing research and related programs, grants for the construction of treatment works, standards and how they are enforced, and permits.

One of the primary objectives in establishing the CWA was to develop national programs for the prevention, reduction, and elimination of pollution. To achieve this objective, research, investigations, and experiments are required to identify ways of preventing future contamination problems. Studies are required to identify the types and distribution of contaminants at waste sites, with the intent of their future reduction and/or elimination. Other studies are striving to better understand the causes and effects that contaminants have on human health and the environment.

This statute requires the federal government to work with individual states and other interested agencies, organizations, and persons to investigate pollution sources. Research fellowships at public or nonprofit private educational institutions or research organizations are required to be made available to conduct research on the harmful effects on the health and welfare of persons who are exposed to pollutants in water. Grants must also be made available to any state agency or interstate agency to assist projects which focus on developing new or improved methods of waste treatment or water purification. Water bodies that are specifically addressed in the CWA include the Hudson River,

Chesapeake Bay, Great Lakes, Long Island Sound, Lake Champlain, and Onondaga Lake (ELR 1990).

Grants from the federal government are provided for the construction of treatment works to assist the development and implementation of waste treatment management plans and practices which help to achieve the goals of the CWA. These plans and practices are expected to use the best practicable waste treatment technologies available. To the extent practicable, waste treatment management is to be maintained on an areawide basis, to provide control or treatment of all point and nonpoint sources of pollution. Preference is always placed on waste treatment management which results in the construction of revenue-producing facilities providing for the:

- recycling of potential sewage pollutants through the production of agriculture, silviculture, or aquaculture products
- confined and contained disposal of pollutants not recycled
- reclamation of wastewater, and
- ultimate disposal of sludge in a manner that will not result in environmental hazards.

The statute requires preference to be placed on waste treatment management which results in integrating facilities for sewage treatment and recycling with facilities to treat, dispose of, or utilize other industrial and municipal wastes such as solid waste, waste heat, and thermal discharges (ELR 1990).

The statute seeks to control pollutants being discharged into water by two permit programs. The first is the National Pollutant Discharge Elimination System. Either EPA, or a state which has established its own program, issues permits for the discharge of any pollutant or combination of pollutants on condition that the discharge will meet all applicable standards or requirements. The U.S. Army Corps of Engineers issues another type of permit under the CWA for discharging dredged or fill material into navigable waters.

The CWA also defines standards for measuring pollution in water. Unless a source discharges into a publicly owned treatment works, point source effluents must achieve the best practicable control technology and must satisfy pretreatment requirements for toxic substances. In addition, water quality standards apply where the technology-based limitations fail to attain a level protective of the public health. The EPA also periodically issues new source performance standards which must take cost factors and environmental impacts into consideration (ELR 1990).

SDWA Compliance

The purpose of the Safe Drinking Water Act (SDWA, 42 USC §§300f-300j-26) is to protect supplies of public drinking water from contamination. Under this act, the EPA is required to publish maximum contaminant level goals (MCLGs), and to promulgate national primary drinking water regulations (NPDWRs) for certain contaminants in the public water system. Public water systems are either community water systems (have at least 15 connections or regularly serve at least 25 people) or noncommunity water systems (all others). The MCLG is set at the level at which there are no known or anticipated adverse human health effects, in addition to allowing for a margin of safety.

Since the MCLGs are in many cases not attainable, the NPDWR are required to specify maximum contaminant levels (MCLs) that are as close as feasible to the MCLG using the best available technology. The EPA may promulgate a NPDWR that requires the use of a treatment technique in lieu of setting an MCL if it finds that it is not economically or technologically feasible to ascertain the level of the contaminant.

A state has the primary enforcement responsibility for public water systems if the EPA Administrator determines that the state has:

- adopted drinking water regulations that are no less stringent than the NPDWRs
- adequate enforcement and record keeping mechanisms
- provisions for variances and exemptions that conform to the requirements of the statute, and
- an adequate emergency drinking water plan.

The statute authorizes courts to issue injunctions and assess civil penalties against violators unless they have an approved variance. Variances may be granted to public water systems if they have implemented the best available technology or treatment technique but cannot meet the applicable MCLs requirements because of characteristics of reasonably available raw water sources (ELR 1990).

After June 19, 1986, this statute requires that any pipe, solder, or flux used in the installation or repair of any public water system, or any plumbing providing water for human consumption that is connected to a public water system is required to be free of lead. In addition, each state is required to assist local educational agencies in testing for and remedying lead contamination in school drinking water.

The statute protects underground sources of drinking water by preventing underground injection that may endanger a drinking water source. To further protect underground water sources, states are required to establish programs to protect wellhead areas from contamination. Finally, the statute regulates drinking water coolers with lead-lined tanks. Manufacturers and importers of coolers with lead-lined tanks were required to repair, replace, or recall them within one year after October 31, 1988 (ELR 1990).

CAA Compliance

The primary objective of the Clean Air Act (CAA, 42 USC §§7601–7671q) is to protect the quality of our nation's air by:

- requiring air quality monitoring stations in major urban areas and other appropriate areas throughout the United States
- providing daily analysis and reporting of air quality data using a uniform air quality index, and
- providing a record-keeping of all monitoring data for analysis and reporting.

The major issues that the statute addresses are noise pollution, acid deposition control, stratospheric ozone protection, and permitting requirements. The EPA has established an Office of Noise Abatement and Control which studies noise and its effect on public health and welfare. The objective of this group is to:

- identify and classify causes and sources of noise
- project future growth in noise levels in urban areas
- study the psychological and physiological effects of noise on humans, and
- determine the effect noise has on wildlife and property.

The CAA is working to control acid deposition in the 48 contiguous states and the District of Columbia through reducing the allowable emissions of sulfur dioxide, and nitrogen oxides. These reductions are being made in a staged approach where prescribed emission limitations must be met by specific deadlines. The statute encourages energy conservation, use of renewable and clean alternative technologies, and pollution prevention as a long-term strategy (ELR 1990).

In an effort to protect the earth's stratospheric ozone layer, all known ozone depleting substances have been classified to control their use. Class I substances have been determined to have an ozone depletion potential of 0.2 or more, and currently include various forms of chlorofluorocarbons, halons, in addition to carbon tetrachloride, and methyl chloroform. Class II substances include various forms of hydrochlorofluorocarbon and have an ozone depletion potential of less than 0.2. As more information becomes available about the causes of ozone depletion, additional substances will be added to these lists. The only time a substance can be removed from the Class II list is if it is moved under the Class I category. No substances under the Class I list can be removed.

The statute has established a permitting program for the purpose of controlling the releases of damaging chemicals into the air. Permits contain information such as enforceable emission limitations and standards, schedules for compliance, reporting requirements to permitting authorities, and any other information that is necessary to assure compliance with applicable requirements of this act (ELR 1990).

REFERENCES

Environmental Law Reporter (ELR), Statute Administrative Proceeding 001, Statute Binder, Environmental Law Institute, 1990.

Environmental Protection Agency (EPA), Guidance for Conducting Remedial Investigations and Feasibility Studies Under CERCLA, EPA 540/G-89/004, 1-7, 1988.

Office of NEPA Project Assistance (ONPA), NEPA Compliance Guide, I, 131-207, 1988.

CHAPTER 2

Developing a Field Sampling Program

This chapter provides guidance on how to develop an effective field sampling program that is both comprehensive and cost-effective. This guidance is based heavily on the EPA's DQO process outlined in EPA 1987 and EPA 1992. Although the guidance provided is slanted toward CERCLA's RI/FS process, the general approach is also of value to the RCRA RFI program.

The first step in developing a field sampling program is to establish DQOs, which are defined as the qualitative and quantitative statements which specify the quality and quantity of data required to support decisions during remedial response activities. DQOs should be developed as an integral part of the Scoping Stage of the RI/FS process (Figure 1.2), and should be carefully outlined in a Sampling and Analysis Plan (SAP). The DQO process results in a well-thought-out SAP, which outlines the chosen sampling and analysis options, and identifies the confidence levels required to make decisions in the remedial process (EPA 1987).

DQOs are applicable to all collection activities, including those performed for a Preliminary Assessment/Site Investigation (PA/SI), Remedial Investigation (RI), Feasibility Study (FS), Remedial Design (RD), and Remedial Action (RA). Since the variability of site characteristics does not allow a generic set of DQOs to address all CERCLA activities, they must be developed on a site-specific basis. When developing DQOs for a specific site, the EPA recommends the following three-stage process (Figure 2.1):

- Stage 1: identification of decision types
- Stage 2: identification of data uses and needs, and
- Stage 3: design of a data collection program.

Stage 1 of the process defines the types of decisions to be made with the investigation results. This is accomplished by identifying and involving the data users, evaluating all of the available data, developing a conceptual model, and specifying the objectives of the project. The conceptual model should describe the suspected sources of contamination, contaminant pathways, and should identify the potential receptors. The process of developing the conceptual model facilitates the identification of decision points, and identifies data insufficiencies (EPA 1987).

The criteria required for determining data adequacy is identified in Stage 2.

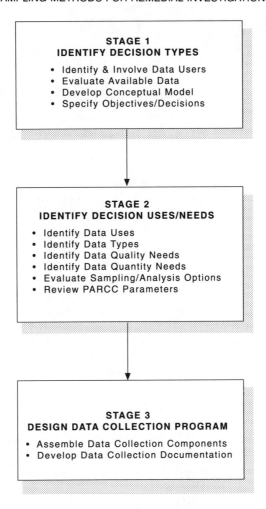

Figure 2.1.   Three stages of the DQO process (EPA 1987).

This stage involves specifying the data needed to meet the Stage 1 objectives, and selecting the sampling and analytical approach that will be both successful and cost-effective (EPA 1987). The specific steps in this stage include: identifying data uses, data types, data quality and data quantity; evaluating sampling and analysis options; and addressing the precision, accuracy, representativeness, completeness, and comparability (PARCC) parameters.

Stage 3 involves the assembly of data collection components, and the writing of a Sampling and Analysis Plan (SAP) which provides all of the details required to perform the fieldwork. The SAP should be composed of a Quality Assurance Project Plan (QAPP) and Field Sampling Plan (FSP) (EPA 1988).

After evaluating the results from the initial phase of sampling, the data adequacy should be reevaluated by repeating Stage 1. If it is determined that additional data are needed, one should repeat Stage 2 and Stage 3. If the data were determined to be sufficient to complete the characterization, one should proceed to the writing of the Feasibility Study (Figure 2.2).

DQO PROCESS STAGE 1

Stage 1 of the DQO process is performed as part of the RI/FS Scoping Effort (Figure 2.2). The major objectives of this stage are to identify the data user, obtain all available information, identify the contaminants of concern, assess the adequacy of the current database, develop a site conceptual model, and identify the objectives of the required investigations (Figure 2.3).

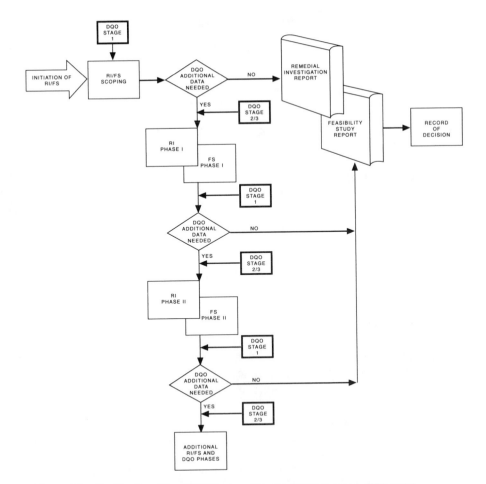

Figure 2.2.   Positioning of the DQO Stages within the RI/FS Program (EPA 1987).

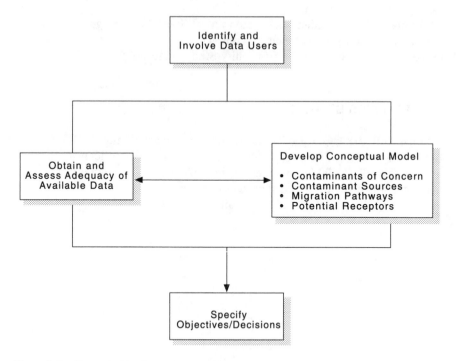

Figure 2.3.   Stage 1 objectives of the DQO process (Modified after EPA 1987).

Identifying the Data User

Due to the interdisciplinary nature of remedial activities, it is important that the data users be identified so that their input can be used in the development of the FSP. Under the RI/FS program, the principal data users include the EPA's Regional Project Manager (RPM), contractor site manager, and contractor lead technical staff, which commonly includes hydrogeologists, risk assessment specialists, and geotechnical and environmental engineers.

Obtaining Available Information

After identifying the data user, all available information about the site should be reviewed as thoroughly and accurately as possible, since this information in combination with information obtained from the Site Inspection will serve as the database for the RI/FS Scoping. During the review, information should be obtained from site historical records, EPA technical and enforcement files, state and local regulatory agency files, U.S. Geological Survey (USGS) files, and other relevant sources.

Information regarding the quality of historical analytical data is particularly important to collect during the review, and should include:

- procedures used to collect the samples
- sampling tools
- sample preservatives
- sample shipment methods
- holding times
- age of the data
- laboratory analytical methods
- method detection limits
- Quality Assurance procedures, and
- documentation of chain-of-custody.

The primary objectives of the Site Inspection should include (EPA 1987):

- using field screening instruments to establish health and safety requirements for site access
- estimating if any site conditions could pose an imminent risk to public health
- confirming information obtained from record review
- recording observed data missing from record review
- performing an inventory of potential sources of contamination
- obtaining site information that will be useful for fieldwork planning purposes.

If the record review and Site Inspection do not provide the minimum data required to meet the Stage 1 objectives, some limited field surveying and/or sampling can be performed to provide this information.

Identifying Contaminants of Concern

The contaminants of concern at a site are identified by comparing the list of chemicals used currently and historically at the site against federal and state water and/or soil quality standards. Those chemicals which have regulated or proposed standards are the primary contaminants of concern.

The federal water quality standards which should be considered in this evaluation include the RCRA MCL, SDWA MCL, and the SDWA MCLGs. Individual states often have their own water quality standards, which in some cases are more stringent than the federal standards. Since there are currently no federal MCLs for soils, and very few state guidelines, it is recommended that all chemicals identified in soil that have a corresponding water quality standard should be considered contaminants of concern. Preliminary Remediation Goals (PRGs) are commonly calculated for soils, and are based on risk calculations. Also, soil background levels are commonly estimated to aid in determining what chemical concentrations define contamination.

Once the contaminants of concern have been identified, the site background information should be revisited to identify which areas of the site require characterization to determine the nature and extent of contamination.

Assessing Adequacy of the Current Database

After all of the information gathered from the record review and Site Inspection has been compiled, an assessment is made of the data adequacy. This assessment is performed by first validating the usability of the data, then plotting the data on a site map so that the location and distribution of the data points can be evaluated. With these data, an attempt should be made to draw the aerial extent of contamination on a map based on specific action levels. This exercise will make evident those areas where additional data are needed. A statistical evaluation of the data set will determine the actual number of data points needed to meet the negotiated probability requirements (see DQO Process Stage 2, below).

Developing Site Conceptual Model

After the data adequacy assessment is complete, a site conceptual model can be developed. The model describes the site and the environments that it contains, and presents hypotheses regarding the contaminants present, their routes of migration, and their potential impact on sensitive receptors. The model should be detailed enough to address the potential or suspected sources, types, and concentrations of contaminants, affected media, rates and pathways of migration, and receptors (EPA 1987). The basic elements of the model include the:

- source
- pathway, and
- receptors.

To adequately define the source of contamination, one needs to identify the specific types and concentrations of contaminants present, and the boundaries of the contaminant plume. Some of the more common sources of contamination include leaking aboveground or underground tanks, drums, transfer lines, and landfills.

The primary pathways for contaminant migration are through the air, surface water, groundwater, and through the food chain (Figure 2.4). Air can be a significant pathway when the contaminants of concern include volatile organics, acids, radionuclides, and/or biological hazards such as viruses. This pathway is of most concern when contamination is located near the ground surface. The fumes derived from acids and volatilized organic compounds can be rapidly transported through the air pathway to downwind receptors. Other concerns to the air pathway include radiological contaminants in the form of radon gas or alpha particles which attach themselves to dust particles, as can biological viruses.

Surface water generated from rain or onsite activities provides a pathway for rapid contaminant migration, and can carry contaminants either in solu-

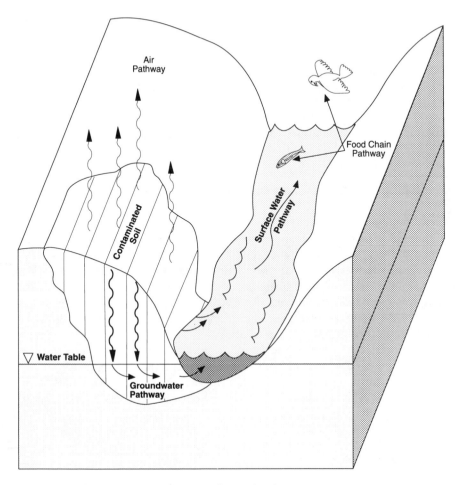

Figure 2.4. Primary pathways for contaminant migration.

tion or in particulate form. As the surface water drains into sewer lines, creeks, streams, and/or rivers, contaminants are rapidly transported to downstream receptors. Similarly, groundwater can carry contaminants in solution or in particulate form. However, the transport velocities for groundwater are typically much slower than for surface water.

Contaminants can also migrate through the food chain. For example, contaminants in soil can often be taken up by vegetation growing in the area. Through the consumption of this vegetation, herbivorous animals receive a dose of the contamination. This contamination is then passed on to the carnivore.

Receptors are those humans or animals which are exposed to the contamination either through inhalation, dermal contact, or ingestion. The dose received by the receptor is dependent on the frequency and degree of exposure. For example, it is obvious that a person who works 8 hr in a contaminated area receives a higher dose than one who works only 4 hr performing the same task. However, if the tasks vary, it is possible for the person working only 4 hr to receive the higher dose.

To assist the investigator in the development of the conceptual model, various types of groundwater, air quality, and geostatistical computer modeling should be considered. This modeling can be used to guide the data collection program. Sensitivity analyses can help identify the types of data needed, as well as critical sampling locations.

Developing Future Sampling Objectives

The final step under Stage 1 is to specify the objectives of future field sampling efforts in a clear and precise decision statement. These objectives should address the major areas of the remedial process, which include characterizing the site with respect to the environmental setting, proximity, and size of the human population, and nature of the environmental problem. These objectives should identify potential remedies, determine specific performance levels for the potential remedies, and evaluate the consequences of making a wrong decision regarding site remediation.

The primary objectives of the sampling program should be to identify the source(s) and aerial extent of contamination at the site, as well as to collect data which can be used to predict the rate of contaminant migration. This information is critical since it will ultimately be used to assist the selection of the remedial alternative.

DQO PROCESS STAGE 2

Stage 2 of the DQO process defines data uses and specifies the types of data needed to meet the project objectives. This stage begins after the conceptual model and project objectives have been established. The primary elements of this stage involve (Figure 2.5):

- identifying data uses and data types
- identifying data quality and quantity needs
- evaluating sampling and analysis options, and
- reviewing Precision, Accuracy, Representativeness, Completeness, and Comparability (PARCC) parameters.

Identifying Data Uses and Data Types

The EPA recommends that project managers use a form, such as that presented in Figure 2.6, to assist in defining data uses (EPA 1987). The most applicable data use categories include:

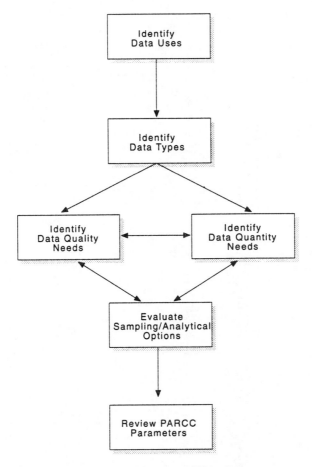

Figure 2.5. Stage 2 objectives of the DQO process (EPA 1987).

- site characterization
- health and safety
- risk assessment
- evaluation of alternatives
- engineering design of alternatives
- monitoring during remedial action, and
- identifying the Potentially Responsible Party (PRP).

Once all of the data uses have been identified, they need to be prioritized based on the most demanding use of each type of data. The data quality required will be based on the acceptable limits of uncertainty established by decision makers. The limits of uncertainty drive the selection of both the analytical and sampling approaches.

Using the results from the data adequacy assessment completed under Stage 1, the types of data required to complete the characterization should be evi-

DATA USES

Site

Name _____

Location _____   EPA Region _____   Date _____

Number _____   Contractor _____

Phase RI1 RI2 RI3 ERA FS RD RA   Site Manager _____

Data Use / Media	Site Characterization (Including Health & Safety)	Risk Assessment	Evaluation of Alternatives	Engineering Design of Alternatives	Monitoring During Remedial Action	PRP Determination	Other
Source Sampling Type _____							
Soil Sampling							
Ground Water Sampling							
Surface Water/Sediment Sampling							
Air Sampling							
Biological Sampling							
Other _____							

Figure 2.6. Checklist which can be used to assist defining data uses (EPA 1987).

dent. There are a number of media which can be sampled as part of a site characterization; however, sampling of all of these media is not required for all sites. The primary media which are commonly sampled as part of a characterization study include:

- shallow soil
- deep soil
- sediment
- surface water
- groundwater
- bedrock
- soil-gas, and
- ambient air.

For more complicated and more hazardous sites, a number of secondary media can be sampled to further define the extent of contamination, and to assist in the evaluation of the baseline risk that the site poses. These media include, but are not limited to:

- sludge
- soil moisture
- concrete
- paint
- dust
- flora, and
- fauna.

Identifying Data Quality Needs

After the data uses and data types have been identified, one needs to identify the quality of the data required. The term data quality refers primarily to the accuracy and precision of the data collected. The higher the data quality, the more confidence the investigators have in the accuracy of the results.

Very rarely will all the data collected during a field sampling program require the same level of data quality. More commonly, a sampling program collects a large variety of data, each type requiring a different level of quality. For example, in the process of collecting a groundwater sample for laboratory analysis, field screening instruments are used to monitor the air quality for health and safety purposes, and to monitor the water properties to determine when the well has stabilized. The level of data quality required to perform this type of screening is drastically different from the quality required to analyze a groundwater sample to determine if it exceeds drinking water guidelines. The EPA has defined five separate data quality levels (EPA 1987):

- Level I: Data at this level are collected with field screening instruments such as an organic vapor analyzer, explosivity meter, etc. These analytical results are not typically compound specific, but are able to provide results in real-time.
- Level II: Data at this level are collected in the field using more sophisticated field analytical instruments, such as a gas chromatograph setup in a mobile laboratory. The quality of data generated at this level depends on the use of suitable calibration standards, reference materials, and sample preparation equipment. Analytical results at this level are commonly available in real-time or several hours.
- Level III: At this level, all analyses are performed in an offsite analytical

laboratory. Level III analyses commonly use SW-846 or Contract Laboratory Program (CLP) procedures. At this level, it is not necessary to utilize the validation of documentation procedures required for CLP Level IV analyses. The laboratory is not required to be a CLP laboratory.

- Level IV: At this level, all analyses are performed in an offsite CLP analytical laboratory following CLP protocols. This level is characterized by rigorous Quality Assurance/Quality Control (QA/QC) procedures and documentation.
- Level V: At this level, analyses are performed by nonstandard methods in an offsite analytical laboratory. The laboratory does not need to be a CLP laboratory. Method development or modification may be needed to meet required detection limits.

For sites on the NPL, critical sampling points which help define the vertical or horizontal boundaries of contamination should be of Level IV quality. Prior to defining the data quality requirements, one must become familiar with the Applicable or Relevant and Appropriate Requirements (ARARs) for each environmental media. It is critical that detection limits for the analytical methods selected be low enough to detect these levels. It is also critical that all of the components of the remedial action process be considered when setting the data quality requirements.

More recent EPA guidance recommends determining data quality requirements through the establishment of uncertainty constraints, which define what is an acceptable probability of making an incorrect decision. These uncertainty constraints are used to establish quantitative limits on total study error and corresponding measurement and sampling error constraints (EPA 1992).

Uncertainty constraints should be set after carefully evaluating the consequences of making an incorrect conclusion. When making this evaluation, the decision maker should consider political, social, and economic consequences of decision error. A statistician should be used to assist in the setting of uncertainty constraints to ensure that they are feasible and complete.

The steps that are involved in setting acceptable probabilities for decision error include (EPA 1992):

- defining false positive and false negative errors for the decision, and describing the consequences of each type of error,
- evaluating these consequences according to the relative level of concern they would cause, with emphasis on the environment, public health, economics, and social and political consequences,
- determining if false positive or false negative errors are of greater concern,
- establishing, with the assistance of a statistician, an acceptable probability for the occurrence of each of these errors,
- combining the probability statement into a formal statement of the levels of uncertainty that can be tolerated in the results, and
- reviewing the decision rule. If necessary, one may revise or add quantitative measures that will allow the decision uncertainty to be evaluated.

Identifying Data Quantity Needs

In Stages 1 and 2 of the DQO process, the investigator must determine the acceptable probability of not finding an existing contaminated zone. This determination should be made with input from the lead regulatory agency. Based on this determination, the existing database can be evaluated to determine which areas require further characterization. This evaluation should be performed at the completion of each phase of the remedial investigation (Figure 2.2).

The EPA recognizes three general approaches in selecting specific sampling locations. These approaches include (EPA 1989a):

- biased (judgment) sampling
- systematic sampling, and
- random sampling.

Biased sampling involves selecting sampling locations based on existing knowledge of the release, such as visual evidence of contamination in the soil, or knowing that the contamination was carried down a particular drainage. A few examples of biased sampling include (EPA 1989a):

- collecting air samples downwind of an area known to be contaminated
- collecting surface water and sediment samples from a drainage channel that receives surface water runoff from a known contaminated area, and
- collecting groundwater samples from locations downgradient from a known contaminated area.

Biased sampling tends to bias the data obtained toward higher contaminant concentrations, since samples are selected from areas either known or suspected to be contaminated.

The systematic and random sampling approach involves collecting samples from locations established by a predetermined sampling scheme, such as a line or a grid. Samples collected with this approach are unbiased, and can be used for modeling or calculating the average concentration of contaminants at a site. When selecting the size of the grid, one should consider the following:

- the gridded area should be larger than the suspected extent of contamination
- while the shape of the grid cells should be square, the overall grid can be of any shape
- the size of the grid cells should be based on confidence level requirements negotiated with the lead regulatory agency. The smaller the grid cell, the higher the confidence that a contaminant area will not be overlooked.

To minimize sampling bias, the exact sampling location within each cell of the sampling grid can be chosen systematically or randomly. When systematic sampling, samples are collected at grid intersections. Random sampling involves using a "randomizing scheme," such as a random number table to select the sampling location within each grid cell. Random sampling is most commonly used at sites where the spatial distribution of the contaminants is expected to be highly variable. Regardless of which sampling approach is selected, it is recommended that the time be taken to mark the grid intersections in the field to assure that the sampling points are located accurately (EPA 1989a).

Figure 2.7 illustrates a systematic and random sampling pattern. A possible limitation to the systematic sampling approach is that if the contaminants are distributed in a regular pattern, the sampling points could all fall within the clean area.

Typically biased sampling is performed around areas known or strongly suspected to be contaminated, to define the extent of contamination, while the systematic and random sampling approach is commonly performed in areas where little information is known. When contamination is identified at a sampling point within a random or systematic sampling scheme, additional sampling should be performed around that point to define the extent of contamination. This can be accomplished using the biased sampling approach, or the random or systematic approach using a tighter grid spacing.

The grid spacing used for systematic or random sampling should be tight enough to meet the probability requirements that a contaminant plume will

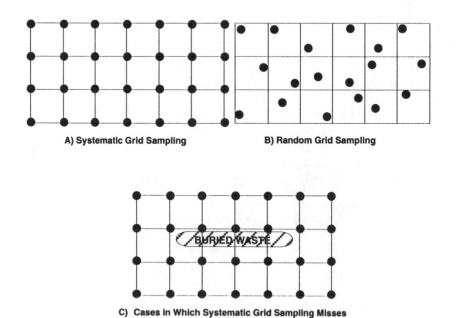

A) Systematic Grid Sampling   **B) Random Grid Sampling**

**C) Cases in Which Systematic Grid Sampling Misses
Wastes Buried in a Regular Pattern**

Figure 2.7. Example of a systematic versus a random sampling scheme (Modified after EPA 1989A).

not be overlooked. When determining the appropriate grid spacing, two assumptions are required (EPA 1987):

- the approximate size and shape of the contaminated zone (target) must be known, and
- any sample collected within the zone of contamination will identify the contamination.

If the samples are collected in a perfectly regular grid and the target is circular, the probability of hitting the target for a given grid size is as shown in Table 2.1 (Gilbert 1982).

If the target is not circular, a simulation procedure must be used, which uses a "hit" or "miss" simulation involving the following steps (EPA 1987):

- simulate the shape of the target
- randomly locate the target within the site
- determine if any sampling locations from a proposed grid spacing fall within the boundaries of the target. If so, score a "hit," otherwise a "miss"
- simulate and randomly locate several hundred targets using a computer program and record the number of hits and misses.

To calculate the probability of locating the contaminated zone, take the total number of hits and divide by the total number of misses. By varying the number of samples for a fixed target, the number of samples required to lower the risk of missing the contamination to an acceptable level can be determined. This method allows one to determine both the number and location of samples necessary to satisfy the DQOs.

For further guidance on using statistics in the development of sampling strategies, see Gilbert 1982 and 1987, and EPA 1987 and 1992.

Evaluating Sampling and Analysis Options

A field sampling program should optimumly be performed as shown in Figure 2.8, where initial site characterization is performed with a heavy emphasis on using various types of screening technologies (Level I and II). Level I screening is performed onsite using hand-held instruments, such as an organic vapor analyzer, pH meter, and explosivity meter. Level II screening is also typically performed onsite and includes techniques such as soil-gas surveying, surface and downhole geophysics, and cone penetrometry. As more information becomes available about the source and extent of contamination,

Table 2.1. Probability of Hitting the Target for a Given Grid Size

Probability	Grid Spacing/Diameter of Target
0.8	1.13
0.9	1.01
0.95	0.94
0.99	0.86

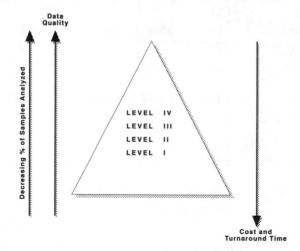

Figure 2.8. Optimum use of remedial investigation dollars (Modified after EPA 1987).

samples are collected for specific laboratory analysis (Level III and/or Level IV). Collecting data in this manner is most cost-effective.

Reviewing PARCC Parameters

PARCC parameters are all indicators of data quality. In an ideal situation, numerical precision, accuracy, and completeness goals would be established and these goals would aid in selecting the most appropriate analytical technique. However, since environmental sites are so different, it is not practical to set universal PARCC goals. Rather, the EPA recommends that historical precision and accuracy achieved by different analytical techniques should be reviewed to aid in selecting the most appropriate analytical technique (EPA 1987).

Precision is defined as a measure of the size of the closeness of agreement among individual measurements, while accuracy is defined as a measure of the closeness of measurements to the true value (Gilbert 1987). Precision and accuracy are both affected by sampling and analytical factors. The analytical effect on precision and accuracy is more easily controlled than the sampling effect due to the environment. Some of the more common ways that a sample can be contaminated during the sampling process include:

- using an inappropriate sampling tool
- using a sampling tool made from an inappropriate material
- using a contaminated sampling tool
- using a contaminated sample bottle, and
- improperly preserving a sample bottle.

The primary method of controlling field sampling contamination is through the collection and analysis of quality control samples such as replicates, rinsate blanks, field blanks, and trip blanks. Similarly, laboratories analyze blank and spiked samples as a quality control mechanism.

Representativeness reflects the degree to which sample data accurately and precisely represent a characteristic of a population, parameter variations at a sampling point, or an environmental condition (EPA 1987). Representativeness is a qualitative parameter which is primarily concerned with the proper design of the sampling program. This criterion is most appropriately satisfied by being certain that a sufficient number of samples are collected, and the sampling locations are carefully positioned at representative locations.

Completeness is a term that refers to the percentage of measurements which are determined to be valid. CLP data has been found to be 80% to 85% complete on a nationwide basis. This can be extrapolated to indicate that Level III, IV, and V analytical techniques will generate data that are approximately 80% complete. Level I and II data should be expected to have a lower completeness value due to the difficulties associated with collecting measurements in the field (EPA 1987).

Comparability is a qualitative parameter expressing the confidence with which one data set can be compared with another. This parameter is limited by the other PARCC parameters since data sets can be compared with confidence only when precision and accuracy are known (EPA 1987).

DQO PROCESS STAGE 3

The last stage in the DQO process involves the design and development of a data collection program which is presented in a SAP. The SAP is composed of a QAPP and FSP (EPA 1988). The purpose of the QAPP is to describe the policy, organization, functional activities, and quality assurance and quality control requirements necessary to meet the DQOs. The QAPP should address the following primary elements (EPA 1987):

- sampling procedures
- sample and document custody procedures
- calibration procedures and frequency
- analytical procedures
- data reduction, validation, and reporting
- internal quality control checks
- performance and system audits
- preventive maintenance
- data measurement assessment procedures
- corrective actions, and
- quality assurance reports to management.

On the other hand, the FSP should provide the following information:

- site background information
- summary of contaminants of concern

- summary of DQOs
- maps showing sampling locations
- rationale for selected sampling points
- text describing sampling tools and methods
- tables outlining analyses to be performed
- tables presenting bottle and preservation requirements
- tables presenting analytical holding times
- tables presenting analytical methods and required detection limits
- details on required documentation
- description of sample numbering system
- field sampling procedures
- decontamination procedures, and
- details on how to handle investigation-derived waste.

To make the FSP more implementable in the field, it should be written as a short and concise document, and should utilize summary tables whenever possible. A recommended general outline for an FSP is presented in Figure 2.9. If fieldwork is to be performed in several phases, a separate or modified SAP will be required for each phase. The term "phase" refers to a segment of fieldwork separated from other phases by enough time to require remobilization of sampling equipment and personnel. Each phase of sampling is commonly performed in several stages, where the results from the first stage of sampling are used to decide whether or not to proceed to the next stage. This method of sampling is often referred to as the "observational approach." These stages are not always performed consecutively.

For example, during Phase I of a field sampling program, Stage 1 sampling activities may involve collecting an initial set of groundwater samples using the Direct Push Method, downgradient from a site, to determine if contamination is migrating offsite. If groundwater contamination is identified during Stage 1 in concentrations exceeding regulatory guidelines, work would proceed to Stage 2, where further DPM sampling would be performed to better define the boundaries of the contaminant plume, prior to moving to Stage 3, where groundwater monitoring wells are installed. On the other hand, if contaminants are detected in groundwater but not in concentrations exceeding regulatory guidelines, work may proceed directly to Stage 3 where long-term monitoring wells are installed.

Primary Characterization Tools

One of the most critical and challenging aspects of developing a field sampling program is selecting sampling equipment and methods. With all of the sophisticated sampling equipment on the market today, it is important to select a tool that complements the analyses to be performed. If a poor selection is made, the resulting analytical data will not accurately reflect site conditions. The selected method should also be relatively easy to implement and safe.

Before beginning the method selection process, it is important to first decide

1. Introduction

    1.1 Site Background
    1.2 Site Description
    1.3 Summary of Existing Site Conditions
    1.4 Contaminants of Concern
    1.5 Purpose and Scope
    1.6 Data Quality Objectives

2. Field Investigation Approach

    2.1 Phase I

        2.1.1 Stage 1 Sampling

            2.1.1.1 Soil Sampling
            2.1.1.2 Sediment Sampling
            2.1.1.3 Surface Water Sampling
            2.1.1.4 Groundwater Sampling
            2.1.1.5 Air Sampling
            2.1.1.6 Other Sampling

        2.1.2 Stage 2 Sampling

        2.1.3 Stage 3 Sampling

    2.2 Additional Phases

3. Analytical Procedures

    3.1 Quality Control
    3.2 Sample Handling, Packaging, and Shipping

4. Field Notebooks and Documentation

5. Decontamination

6. Handling of Investigation-Derived Waste

7. References

Figure 2.9. Proposed outline for a Field Sampling Plan.

whether a grab, composite, or integrated sample is most appropriate. A grab sample is an individual sample taken at a specific location at a specific time. When a contaminant source is known to vary in composition over time, grab samples collected at suitable intervals and analyzed separately can provide useful information regarding the magnitude and duration of the variations (EPA 1989). Examples of grab sampling methods include:

- using a scoop to transfer a sediment from one location directly into a sample jar
- spooning surface soil from one location directly into a sample jar
- spooning deep soil from a split-spoon sampler directly into a sample jar
- submerging a water sample bottle below the water surface in a pond, and
- using a bailer to collect a groundwater sample from a well, then transferring the water into a sample bottle.

Composite samples are collected by homogenizing a sampling interval, or by homogenizing samples collected from different locations and/or times. The results from composite samples provide average concentrations of contaminants and are generally not effective in defining hot spots. Composite samples are used in some cases to reduce the number of individual samples collected. Compositing can be used to assist in defining the overall extent of a contamination; however, it should not be used to substitute for characterizing individual constituent concentrations. Therefore, compositing should not be overused, and should always be done in conjunction with an adequate number of grab samples (EPA 1989). A few common examples of how composite samples are collected are as follows:

- transferring deep soil from one or more split-spoon samplers into a stainless steel bowl and homogenizing before transferring into sample jars
- collecting several scoops of soil from each of several waste drums and homogenizing before transferring into sample jars, and
- collecting surface water by partially filling a sample bottle from several locations or depths.

An integrated sample is commonly collected by continuously collecting single samples to describe a population in which one or more parameters vary with time. Integrated samples are most often collected from water as opposed to soil media. Time integrated samples can provide average concentrations over the period of time sampled. For example, this method would be effective for characterizing the composition of a process waste stream which varies in composition based on the processes being performed. Integrated sampling devices can be programmed to collect samples at regular intervals or at biased sampling times (EPA 1989).

In general, samples to be analyzed for volatile organics should be collected as grab samples, since the composite and integrated methods provide the opportunity for volatiles to be lost into the atmosphere. Samples for radiological analysis are also preferably collected as grab samples since the contamination commonly occurs in particulate form. Composite samples can effectively be collected for semivolatiles, metals, pesticides/PCBs, total petroleum hydrocarbons (TPH), and other common analyses. If soil composites are collected from a soil boring, the composited interval should not exceed several feet since compositing larger intervals tends to dilute the sample beyond the point of providing useful data. Integrated sampling is more commonly used to collect surface water as opposed to groundwater samples since the chemical composition of surface water typically shows greater variation.

Equipment used to collect samples for laboratory analysis should be made of either Teflon and/or stainless steel. Teflon is preferred over stainless steel when samples are to be analyzed for primary metals. When soil samples are collected for volatile organic analysis, stainless steel sample tube liners can be used with some methods. Since a tube liner can be capped shortly after the sample is collected, there is less opportunity for losing volatile organics to the atmosphere.

Tables 2.2 through 2.8 summarize which methods are most effective for some of the more common analytical tests and procedures. Once the appropriate sampling procedure has been selected, the corresponding SOP provided in the following sections should be followed.

Secondary Characterization Tools

To reduce the high cost and inefficiency of characterizing sites by collecting and analyzing individual environmental samples, secondary characterization techniques such as surface and downhole geophysics, underground pipe surveying, soil-gas, cone penetrometry, and HydroPunch should be considered. Surface geophysical surveying techniques can provide valuable information about soil and bedrock structure, location of buried contaminant sources, and assist in defining the aerial extent of contamination. The most effective geophysical surveying techniques that are currently used for environmental surveying include:

- electromagnetics
- ground-penetrating radar
- magnetics
- radiometrics
- seismic
- gravity, and
- remote sensing.

The effectiveness of each of these surveying techniques is dependent on site-specific soil properties, as well as the presence of local interferences such as overhead or underground power lines.

Downhole geophysical surveying can be used to assist in defining a number of formation properties such as:

- borehole lithology
- bedrock fracture zones
- zones of saturation
- groundwater flow velocity and direction
- electrical properties
- temperature, and
- radiation levels.

Table 2.2. Evaluation Table for Shallow Soil Sampling Methods

	Laboratory Analyses								Sample Type			Sampling Depth		Lithology Description
	Volatiles	Semi-Volatiles	Primary Metals	Pesticides	PCBs	TPH	Radionuclides	Geotechnical	Grab	Composite (Vertical)	Composite (Areal)	Surface (0.0-0.5 ft.)	Shallow (0.5-5.0 ft.)	
Scoop	2	1	1	1	1	1	1		1		1	1		1
Hand Auger		1	1	1	1	1	1			1	2	1	1	1
Slide-Hammer	1	1	1	1	1	1	1		1	1	2	1	1	2
Open-Tube									1			1	1	1
Split-Tube/Solid-Tube	1/1	1/1	1/1	1/1	1/1	1/1	1/1		1/1	1/2	2/2		1	1/2
Thin-Walled Tube								1	1				1	

1 = Preferred Method
2 = Acceptable Method
Empty Cell = Method Is Not Recommended

Table 2.3. Evaluation Table for Deep Soil Sampling Methods

	Laboratory Analyses								Sample Type			Sampling Depth	Lithology Description
	Volatiles	Semi-Volatiles	Primary Metals	Pesticides	PCBs	TPH	Radionuclides	Geotechnical	Grab	Composite (Vertical)	Composite (Areal)	Deep (>5.0 ft.)	
Split-Tube/Solid-Tube	1/1	1/1	1/1	1/1	1/1	1/1	1/1		1/1	1/2	2/2	1/1	1/2
Thin-Walled Tube								1	1			1	

1 = Preferred Method
2 = Acceptable Method
Empty Cell = Method Is Not Recommended

Table 2.4. Evaluation Table for Stream, River, and Surface Water Drainage Sediment Sampling Methods

	Laboratory Analyses								Sample Type			Sampling Depth		Lithology Description
	Volatiles	Semi-Volatiles	Primary Metals	Pesticides	PCBs	TPH	Radionuclides	Geotechnical	Grab	Composite (Vertical)	Composite (Areal)	Surface (0.0-0.5 ft.)	Shallow (0.0-3.0 ft.)	
Scoop or Dipper	2	2	2	2	2	2	2		2		2	2		2
Slide-Hammer	1	1	1	1	1	1	1		1	1	1	2	1	2
Box Sampler	2	1	1	1	1	1	1		1		1	1		1

1 = Preferred Method
2 = Acceptable Method
Empty Cell = Method Is Not Recommended

Table 2.5. Evaluation Table for Pond, Lake, and Retention Pond Sediment Sampling Methods

	Laboratory Analyses								Sample Type			Sampling Depth		Lithology Description
	Volatiles	Semi-Volatiles	Primary Metals	Pesticides	PCBs	TPH	Radionuclides	Geotechnical	Grab	Composite (Vertical)	Composite (Areal)	Surface (0.0-0.5 ft.)	Shallow (0.0-3.0 ft.)	
Scoop or Dipper	2	2	2	2	2	2	2		2		2	2		2
Slide-Hammer	1	1	1	1	1	1	1		1	1	1		1	1
Box Sampler	2	1	1	1	1	1	1		1		1	1		1
Dredge Sampler	2	1	1	1	1	1	1		1		1	1		1

1 = Preferred Method
2 = Acceptable Method
Empty Cell = Method Is Not Recommended

Table 2.6. Evaluation Table for Stream, River, and Drainage Surface Water Sampling Methods

	Laboratory Analyses							Sample Type				Sampling Depth		
	Volatiles	Semi-Volatiles	Primary Metals	Pesticides	PCBs	TPH	Radionuclides	Grab	Composite (Vertical)	Composite (Areal)	Integrated	Surface (0.0-0.5 ft.)	Shallow (0.5-5.0 ft.)	Deep (>5.0 ft.)
Bottle Submersion	1	1	1	1	1	1	1	1		1	1	1		
Dipper	2	1	1	1	1	1	1	1		1	1	1		
Extendable Bottle Sampler	2	1	1	1	1	1	1	1	1	1	1	2	1	
Extendable Tube Sampler	2	1	1	1	1	1	1	1	1	1	1	2	1	

1 = Preferred Method
2 = Acceptable Method
Empty Cell = Method Is Not Recommended

Table 2.7. Evaluation Table for Pond, Lake, and Retention Pond Surface Water Sampling Methods

	Laboratory Analyses							Sample Type				Sampling Depth		
	Volatiles	Semi-Volatiles	Primary Metals	Pesticides	PCBs	TPH	Radionuclides	Grab	Composite (Vertical)	Composite (Areal)	Integrated	Surface (0.0-0.5 ft.)	Shallow (0.0-5.0 ft.)	Deep (> 5.0 ft.)
Bottle Submersion	1	1	1	1	1	1	1	1		1	1	1		
Dipper	2	1	1	1	1	1	1	1		1	1	1		
Extendable Bottle Sampler	2	1	1	1	1	1	1	1	1	1	1	2	1	
Extendable Tube Sampler	2	1	1	1	1	1	1	1	1	1	1	2	1	
Bailer	2	1	1	1	1	1	1	1		1	1	1	2*	
Kemmerer Bottle	2	1	1	1	1	1	1	1	1	1	1		2	1
Bomb Sampler	2	1	1	1	1	1	1	1	1	1	1		2	1

*Able to collect a sample to a depth equal to the length of the bailer.
1 = Preferred Method
2 = Acceptable Method
Empty Cell = Method Is Not Recommended

Table 2.8. Evaluation Table for Groundwater Sampling Methods

	Laboratory Analyses							Sample Type			Sampling Depth	
	Volatiles	Semi-Volatiles	Primary Metals	Pesticides	PCBs	TPH	Radionuclides	Grab	Composite (Vertical)	Integrated	Shallow (0.0-30 ft.)	Deep (>30 ft.)
Bailer	2	1	1	1	1	1	1	1		2	1	2
Bomb Sampler	2	1	1	1	1	1	1	1	1	2	1	2
Bladder Pump	1	1	1	1	1	1	1	1	2	1	1	1
Piston Pump	2	1	1	1	1	1	1	1	2	1	1	1
Submersible Pump	2	1	1	1	1	1	1	1	2	1	1	1

1 = Preferred Method
2 = Acceptable Method
Empty Cell = Method is Not Recommended

These types of surveys provide information which can be used to more clearly define the hydrogeological conditions at the site. This information is particularly useful when evaluating contaminant migration pathways. Some of the more effective downhole geophysical logging techniques which can be used to gather this type of data include:

- caliper
- resistivity
- spontaneous potential
- radar
- induction
- flowmeter
- active gamma
- passive gamma
- neutron
- acoustics
- vertical seismic profiling
- temperature, and
- video.

Underground pipe surveying tools can be used to determine the presence or absence of contamination in buried pipes, drainage lines, and other confined spaces which are either too small or too dangerous to characterize with conventional methods. The surveying tools are either pushed through the pipe with a long cable or are mounted on a remote controlled pipe crawling vehicle. The most effective surveying instruments which are available for pipe surveying include a video camera, organic vapor analyzer, and various types of radiation detection devices.

Soil-gas surveying is commonly used to assist in identifying the source(s) and/or boundaries of volatile organic contamination. This surveying technique involves either performing direct measurements of soil-gas concentrations in the field, or collecting samples in bottles for laboratory analysis.

Cone penetrometry and Direct Push methods can be used to collect soil lithology and groundwater data without needing to drill boreholes or install monitoring wells. These techniques provide the benefit of producing little or no investigation-derived waste (IDW), and can provide data in the field, shortly after sample collection.

Other secondary screening techniques utilize a photoionization detector (PID), flame ionization detector (FID), mercury detector, alpha meter, beta/gamma meter, and downhole gamma detector. A PID or FID is commonly used to screen soil samples for organic vapors. This screening can be performed by either passing the probe of the instrument slowly over the sample immediately after the split-spoon sampler is opened, or more preferably, a small portion of the sample can be placed in a sample jar and allowed to volatilize for several minutes prior to collecting a headspace reading. These same two procedures can be used to screen soil with the mercury detector. The results from these screens are reported in parts-per-million (ppm).

If the contaminants at the site include radionuclides, the probe of an alpha or beta/gamma meter can be passed slowly over the soil in the split-spoon sampler to identify contaminated intervals. The other option for defining radiologically contaminated intervals is to run a downhole gamma survey throughout the depth of the borehole. The results from these surveys are reported in counts-per-minute (cpm).

It is highly recommended that all of the above supplemental characterization tools be considered when developing an FSP since they are very cost-effective and provide valuable data. Whenever possible, confirmation samples should be collected for laboratory analysis to confirm the results from the secondary sampling techniques.

SELECTING SIZE OF SAMPLING TEAM

The optimum size of the sampling team is much dependent on the size of the sampling program. For a small sampling effort performed at a site where the contaminant levels are known or suspected to be low, a team as small as three people is often sufficient. More commonly, a team of four is used, where two are needed for sample collection, labeling, and documentation; a third for health and safety, and quality control; and a fourth for waste management and equipment decontamination.

For larger sampling programs where contaminant levels are relatively low, it is commonly more cost-effective and time-efficient to utilize a much larger sampling team. For example, a 10-member team could effectively be used in the following way:

- six of the members could be broken into two sampling teams of three, each collecting samples simultaneously. Two of the members on each team would be used for sample collection, labeling, and documentation, and the third for health and safety
- two would be used for equipment decontamination, sample preparation, and shipment
- one would be responsible for quality assurance, and
- the last member would be responsible for coordinating all the activities.

If a team is not of sufficient size to support two sampling teams, using additional workers to take care of equipment decontamination, sample preparation, and shipment, can greatly increase the sample collection rate for a single sampling team.

In areas where contamination levels are high, a sampling team of 15 to 20 people is not uncommon. Extra support is needed in these environments since work crews are commonly rotated on a schedule to reduce worker exposure and avoid heat stress. Additional support is also needed to assist in personnel and equipment decontamination. As an example, a 17-member team could effectively be used in the following way:

- eight of the members could be broken into two sampling teams of four which work on rotation. Three of the members on each team would be used for sample collection, labeling, and documentation, and the fourth for health and safety
- two would be used in the contamination reduction zone to decontaminate personnel and equipment exiting the exclusion zone (see Figure 6.1)
- two would be used for equipment decontamination, sample preparation, and shipment
- two would be used for emergency response
- one would be responsible for coordinating the overall health and safety for the site
- one would be responsible for quality assurance, and
- the last member would be responsible for coordinating all of the activities.

If a drill rig were used to assist the sampling, the driller and support personnel would be in addition to the 17-member team outlined above.

FIELD QUALITY ASSURANCE PROCEDURES

To assure that the data collected in the field are representative and valid, a number of quality control checks should be performed. Examples of these checks include (EPA 1989):

- the use of standardized field sampling forms
- verification of sampling data by an independent person
- strict adherence to chain-of-custody procedures
- documenting instrument calibration
- collection of replicate, field blanks, and equipment rinsate samples, and
- use of trip blanks when appropriate.

Using standardized field sampling forms can greatly reduce the amount of writing required in the field, which in turn tends to reduce documentation errors. Another effective way of reducing documentation errors is by using an independent reviewer to check forms for accuracy. It is recommended that an approval block be placed at the bottom of each sampling form, and the reviewer be required to sign the form at the completion of the review.

Immediately following sample collection and preservation, a chain-of-custody seal should be placed over the bottle cap and the bottle placed in an ice-packed cooler. The seal allows the sampler to recognize whether the bottle has been disturbed after the time of collection. Prior to the time that sample custody is signed over to the shipping company, the sampler is responsible for either keeping the sample secured in a locked area or the sample must be kept in sight.

All instruments used in the field must be calibrated prior to use, and the calibration must be documented in a logbook. The recorded information should include at a minimum the:

- calibration time and date
- instrument identification number
- calibration standard used
- whether the calibration was successful or not, and
- calibrator's signature.

To determine the representativeness of the environmental samples collected in the field, quality assurance samples in the form of duplicate samples, field blanks, equipment rinsate blanks, and trip blanks must be collected.

Duplicate samples are collected in the field for the purpose of checking the precision of the laboratory analyses. When collecting a duplicate, it is important that the sample and its duplicate be collected from the same sampling interval at the same time, as opposed to one after the other. For example, when soil or sediment sampling, a duplicate is most effectively collected by first compositing the sampling interval in a stainless steel bowl, then alternately spooning the sample into two separate jars.

Similarly when collecting a duplicate of a groundwater sample, it is preferred that a portion of one sample bottle be filled, then a portion of the duplicate. This procedure is repeated until both sample bottles are full. The only exception to this is when collecting water for volatile organic analysis. In this instance the sample bottles should be filled completely one after the other, since alternately filling these bottles will cause the loss of volatiles. A duplicate surface water sample is most easily collected by simultaneously submerging two sample bottles below the water surface. At a minimum, one duplicate should be collected for every 20 samples collected.

Field blanks are collected for the purpose of determining whether any of the water used from field operations may be providing contamination to the samples being collected. One field blank should be collected from each of the various types of water used during a field investigation, and should be analyzed for all of the parameters of concern at the site. For example, one field blank should be collected from:

- each lot number of distilled/deionized water used for sampling equipment decontamination
- each tap water or other water source used for equipment decontamination, and
- any water source used to assist well development procedures.

Equipment blanks are collected for the purpose of determining the effectiveness of the decontamination procedure. One equipment blank should be prepared each morning prior to the commencement of fieldwork, from each set of sampling equipment to be used that day. For example, if deep soil sampling, shallow soil sampling, and surface water sampling all in the same day, a total of three equipment blanks should be prepared. These blanks are prepared by pouring distilled/deionized water over each set of sampling equipment and catching the rinsate in sample bottles. These blanks need to be analyzed for the same parameter as the samples being collected that day.

Finally, a trip blank is required to accompany each sample shipment where one or more of the samples are being analyzed for volatile organics. The purpose of the trip blank is to determine whether samples were contaminated with volatile organics during the sample shipment. These blanks are prepared in the laboratory by filling a 40-mL volatile organic analysis (VOA) bottle with distilled/deionized water. The laboratory typically sends a box of trip blanks along with each shipment of clean sample bottles.

REFERENCES

Environmental Protection Agency, Characterization of Heterogeneous Wastes, EPA/600/R-92/033, pp. 22–52, 1992.

Environmental Protection Agency, Data Quality Objectives for Remedial Response Activities, EPA/540/G-87/003, Sections 1–5, Appendix A and C, 1987.

Environmental Protection Agency, Extraction and Analysis of Priority Pollutants in Biological Tissue, Method PPB.10/80, S&A Division Region IV, Athens, GA: Laboratory Services Branch, 1980a.

Environmental Protection Agency, Guidance for Conducting Remedial Investigations and Feasibility Studies Under CERCLA, EPA/540/G-89/004, pp. 2–16, 1988.

Environmental Protection Agency, Guidelines for Air Quality Maintenance Planning and Analysis, Volume 10 (revised): Procedures for Evaluating Air Quality Impact of

New Stationary Sources, Research Triangle Park, North Carolina: Office of Air Quality Planning and Standards, 1977.

Environmental Protection Agency, Interim Methods for the Sampling and Analysis of Priority Pollutants in Sediment and Fish Tissue, Cincinnati, OH, ESML, October 1980b.

Environmental Protection Agency, Methods for Measuring the Acute Toxicity of Effluents to Freshwater and Marine Organisms, 3rd Edition, EPA 600/4-85/013, ORD. Cincinnati, OH, March 1985.

Environmental Protection Agency, RCRA Facility Investigation (RFI) Guidance Volume II and III, PB89-200299, 9-70 to 9-72, 12-1 to 12-133, 1989a.

Environmental Protection Agency, RCRA Ground-Water Monitoring: Draft Technical Guidance, PB93-139350, 4-3 to 4-7, 1992.

Environmental Protection Agency, Soil Sampling Quality Assurance User's Guide, EPA/600/8-89/046, pp. 45-84, 1989b.

Environmental Protection Agency (Region VIII), Draft Standard Operation Procedures for Field Samplers, Rev. 4, 1992.

Freed, J.R., P.R. Abell, D.A. Dixon, and R.E. Huddleston, Jr., Sampling Protocol for Analysis of Toxic Pollutants in Ambient Water, Bed Sediment, and Fish, Interim Final Report, Versar, Inc., Springfield, VA, for EPA Office of Water Planning and Standards, Washington, DC, 1980.

Gilbert, R.O., Some Statistical Aspects of Finding Hot Spots and Buried Radioactivity in TRAN-STAT: Statistics for Environmental Studies, No. 19, Pacific Northwest Laboratory, Richland, WA, PNL-SA-10274, 1982.

Gilbert, R.O., Statistical Methods for Environmental Pollution Monitoring, Van Nostrand Reinhold, New York, 1987, p.12.

Mosby, H.S., Manual of Game Investigational Techniques, U.S. Wildlife Society, 1960.

Platts, W.S., W.F. Megahan, and G.W. Minshall, Methods for Evaluating Stream, Riparian, and Biotic Conditions, Draft, U.S. Department of Agriculture Forest Service General Technical Report, INT-138, Ogden, UT, May 1983.

Radian Corporation, Screening Methods for the Development of Air Toxics Emission Factors, EPA-450/4-91-021, 1991.

BIBLIOGRAPHY

Barcelona, M.J., H.A. Wehrmann, M.R. Schock, M.E. Sievers, and J.R. Karny, Sampling Frequency for Ground-Water Quality Monitoring, EPA Project Summary, EPA/600/S4-89/032, 1989.

Barcelona, M.J., J.A. Helfrich, and E.E. Garske, Verification of Sampling Methods and Selection of Materials for Ground-Water Contamination Studies, in A.G. Collins and A.I. Johnson, Eds., Ground-Water Contamination: Field Methods, ASTM STP 963, American Society for Testing and Materials, Philadelphia, PA, 1988, pp. 221-231.

Environmental Protection Agency, Ground-Water Monitoring Technical Enforcement Guidance Document, OSWER Directive No. 9950.1, 1986.

Environmental Protection Agency, Guidance on Feasibility Studies Under CERCLA, EPA/540/G-85/003, 1985.

Environmental Protection Agency, Guidance on Remedial Investigations Under CER-CLA, EPA/540/G-85/002, 1985.

Environmental Protection Agency, Practical Guide for Groundwater Sampling, EPA/600/2-85/104, 1985.

Environmental Protection Agency, RCRA Inspection Manual, Office of Solid Waste, 1981.

Environmental Protection Agency, Sediment Sampling Quality Assurance Users Guide, EPA/600/4-85/048, 1985.

Environmental Protection Agency, Superfund Remedial Design and Remedial Action Guidance, OSWER Directive 9355.0-4A, 1986.

Environmental Protection Agency, Soil Sampling Quality Assurance Users Guide, NTIS PB84-198621, 1984.

Gillham, R.W., M.J.L. Robin, J.F. Barker, and J.A. Cherry, Ground Water Monitoring and Sample Bias, American Petroleum Institute, API Publication No. 4367, 206, 1983.

Yeskis, D., K. Chiu, S. Meyers, J. Weiss, and T. Bloom, A Field Study of Various Sampling Devices and Their Effects on Volatile Organic Contaminants, Second National Outdoor Action Conference on Aquifer Restoration, Ground-Water Monitoring and Geophysical Methods, NWWA, May 23-26, 1988, pp. 471-479.

CHAPTER 3

Field Investigation Methods

In order for a remedial investigation study to be successful, it is necessary to carefully select the most appropriate sampling method for the job. Initial studies often begin by performing screening type surveys using techniques such as aerial photography, surface geophysics, surface radiological surveying and/ or soil-gas surveying, to highlight those areas where contamination may be a problem. The results from these screening surveys are then used to select the points where environmental samples can most effectively be collected for laboratory analysis. The following section provides guidance on which sampling methods are most effective for various sampling objectives. For guidance on how these techniques fit into the development of a field sampling program, see Chapter 2.

Field investigation methods are generally broken down into two major groups, including Nonintrusive and Intrusive Methods. As the title implies, Nonintrusive methods do not require physical penetration of the ground surface. Examples of both of these method are as follows:

Nonintrusive Methods:
- aerial photography
- surface geophysical surveying, and
- surface radiological surveying.

Intrusive Methods:
- soil-gas surveying
- underground pipe surveying
- shallow and deep soil sampling
- sediment sampling
- surface water sampling
- groundwater sampling, and
- drum sampling.

Since most investigative methods have unique environmental or sampling conditions where they work most effectively, it is important to take these into consideration when making your selection. For example, since ground penetrating radar surveying works most effectively in a sandy soil with a low moisture content, one should consider an alternative geophysical or other method when these conditions are not present. Some of the critical factors which should be considered when selecting the optimum sampling technique include:

- clay content of the soil
- moisture content of the soil
- approximate depth to groundwater
- aquifer characteristics
- contaminants of concern
- analyses to be performed on samples
- type of sample being collected (grab, composite, or integrated), and
- sampling depth.

The following sections discuss many of the advantages and limitations of each of these methods, and present SOPs for many of those which do not require specialized academic training. This book focuses more on intrusive sampling techniques since they are in the area of the author's technical expertise. Consequently, some sections such as Aerial Photography, Surface Geophysical Surveying, and Surface Radiological Surveying are presented in less detail.

NONINTRUSIVE METHODS

Prior to collecting samples of environmental media for onsite or laboratory analysis, serious consideration should be given to performing one or more nonintrusive characterization surveys to assist the positioning of intrusive sampling points. Since nonintrusive methods are relatively inexpensive when compared to the cost of obtaining the same data through the collection and chemical analysis of individual environmental samples, they can often save a project money in the long run. The most effective nonintrusive methods that should be considered when developing an FSP include aerial photography, surface geophysical surveying, and surface radiological surveying.

Aerial Photography

Through careful evaluation of historic and/or recent aerial photographs, it is often possible to identify the boundaries of fill areas, potential surface contaminant migration pathways and, in some cases, contaminant source areas. Fill area boundaries can sometimes be defined by identifying unnatural changes in topographic relief, or by identifying unnatural changes in vegetation patterns.

When both historic and recent aerial photographs are available, a comparison of the surface topography between the two sets of photographs will quickly reveal suspect areas. When looking for changes in vegetation types, color photographs are much more effective than black-and-white.

Aerial photographs can be used to identify surface water drainages through which contaminants may be transported from a site. Identifying these pathways is critical to understanding the risk that a site poses to the surrounding environment, since they are routes for rapid contaminant migration. Historic aerial photographs may identify surface water drainages which are no longer present at the site. Since these historical drainages may have been pathways for contaminant migration in the past, one should consider collecting environmental samples from them as well as current drainages.

Source areas of contamination can sometimes be identified with the assistance of these photographs since vegetation near these areas often shows signs of stress. The effectiveness of aerial photography as a site characterization tool will be dependent on the quality and scale of the available photographs.

Surface Geophysical Surveying

Surface geophysical methods can often provide useful data to assist in the identification of buried drums and tanks, boundaries of fill areas, and in some cases the boundaries of a contamination plume. The most effective geophysical methods for a particular site are dependent on a number of factors including the:

- objectives of the survey
- composition of the soil
- soil moisture content
- depth of penetration required, and
- type of surrounding interferences.

The surface geophysical surveying techniques which have proved to be most effective in the environmental industry include:

- electromagnetics
- ground penetrating radar
- magnetics
- radiometrics
- seismic
- gravity, and
- remote sensing.

Since these techniques are out of the realm of the author's technical expertise, specific details on the theory, advantages, and limitations of each method have not been provided. However, these techniques are extremely valuable to a remedial investigation study, and should not be overlooked.

Surface Radiological Surveying

Surface radiological surveying techniques are commonly used for the purpose of locating areas where radiological contamination is concentrated. These surveys are performed over the ground to define locations where the underlying asphalt, concrete, and/or soil may be contaminated, and on the floors, walls, and ceilings of potentially contaminated buildings. These surveys are commonly followed by confirmation soil and/or surface swipe sampling.

Gamma Walk-Over Survey

One of the most common and most effective surface radiological surveying techniques is the Gamma Walk-Over Survey. This survey is performed by carrying an unshielded gamma scintillation detector with a 2-in. by 2-in. NaI probe over a gridded survey pattern for the purpose of identifying radiological "hotspots."

During the survey, the probe of the instrument is held approximately 0.5-in. or less above the ground surface. One can either collect continuous activity levels as the sampler walks across the sampling grid, or individual activity levels can be collected at grid intersections.

For most parts of the country, background gamma activity levels with this instrument commonly range from 7,000- to 9,000-cpm. An activity level of approximately 10,000- to 12,000-cpm has been shown at many sites to be approximately equivalent to the Department Of Energy (DOE) guideline for thorium-232 of 5-pCi/gm.

After the completion of the Gamma Walk-Over Survey, it is common practice to collect a 2-ft soil core sample from each of the hotspots using the Slide Hammer Method (see page 89). The top 0.5-ft of the sample is typically sent to the laboratory for analysis, while the remainder of the sample is placed in archive. If the activity levels of the top 0.5-ft interval exceed applicable regulatory guidelines, the 0.5- to 1.0-ft interval is removed from the archives, and sent to the laboratory for analysis. This is repeated until the depth of contamination is defined. If the depth of contamination exceeds 2-ft, deeper sampling using the Slide Hammer or Split-Tube Methods (see page 89 and 94) will be required.

Building Surface Survey

Screening tools such as the Eberline Model ESP alpha and beta/gamma detectors are commonly used to assist the location of surface contamination on the floors, walls, and ceilings of buildings. Initial screening surveys are commonly performed by first inspecting those areas in the building that would most likely show elevated activity levels if contaminants were present. For example, in a uranium processing plant, one should begin surveying around the uranium presses and other equipment which came in direct contact with the radioactive materials.

If no contamination is identified during this initial screen, there is no need for additional surveying. On the other hand, if elevated activity levels are detected, time should be taken to lay out a sampling grid pattern over the floor, walls, and ceiling of the building using a grid spacing of approximately 10-ft.

After the grid has been laid out, a direct reading radiation measurement should be collected at each grid intersection, and at the center of each square. At those locations showing the highest activity level, a swipe sample should be collected from within a 10-cm by 10-cm area to determine the removable activity level. These swipe samples are commonly counted in an alpha scintillation detector. The probe of the alpha and beta/gamma detector should then be used at each location showing elevated activity levels to define the extent of contamination. The contaminated areas should finally be circled with a paint pen or permanent ink marker to await remediation.

INTRUSIVE SAMPLING METHODS

Intrusive sampling methods are different from nonintrusive methods in that they require the penetration of the ground or water surface. Examples of these methods include soil-gas surveying, underground pipe surveying, and the sampling of shallow and deep soil, sediment, surface water, and groundwater. As discussed earlier in this chapter, serious consideration should be given to performing one or more nonintrusive surveys prior to the commencement of an intrusive sampling program for the purpose of more effectively selecting intrusive sampling locations. To assure the success of a field sampling program, the following questions should be asked prior to the commencement of fieldwork:

- Have the proposed sampling locations been cleared by local utility?
- Is work being performed on private property? If so, have access agreements been prepared and approved by the landowner?
- Has a secured area been established to store drums containing waste materials?
- Have the potential hazards at the site been thoroughly researched, and is a site-specific Health and Safety Plan in place?
- Is there a qualified Health and Safety Officer onsite with the appropriate monitoring equipment to assure the health and safety of the workers?
- Have field quality assurance procedures been developed?
- Has all the sampling and monitoring equipment been properly calibrated?
- Has all sampling equipment been decontaminated?
- Has the laboratory provided the appropriate sample bottles, sample preservatives, and information on the holding time for each of the analyses to be performed?

It is important to contact local utilities prior to performing any intrusive activities. Most areas have a local "Digger's Hotline" telephone number where one can report the locations of proposed sampling points. The hotline representative will provide a Work Authorization Number and a date on which work can begin. If any sampling points are found to be near a utility line, a hotline representative will contact you prior to the work start date and require that the sampling point be moved. If a utility line is encountered during sampling, the Work Authorization Number protects the sampler from liability.

If work is to be performed on private property, it is a legal requirement to get an Access Agreement signed by the landowner prior to beginning work. This agreement should provide background information on the purpose of the study, information regarding the location and depth of the samples to be collected, equipment to be used, and outline the steps that will be taken to restore the site to its original condition.

If wastewater or soil will be generated from any of the sampling activity, a secured area is needed to store the waste material prior to final disposal. Chapter 7 outlines the specific requirements for this storage area.

To assure the health and safety of the field workers, it is essential that all the potential hazards at the site be thoroughly researched, and a site-specific Health and Safety Plan be developed to address each of these hazards. To further ensure the safety of the workers, a Health and Safety Officer must be present onsite with appropriate calibrated instruments to screen the working atmosphere prior to, and during field activities. Based on the results from an initial screening, the Health and Safety Officer must determine the appropriate level of protection to begin work. If conditions change in the process of sampling, the level of protection must likewise change. The Health and Safety Officer is responsible for seeing that all the workers have the appropriate training, are on an annual medical monitoring program, and follow good health and safety practices while performing fieldwork.

To assure the effectiveness of the field sampling procedures, a Quality Assurance Project Plan must be in place at the time of sampling. The primary objective of this plan is to outline the frequency and method of collecting quality control samples, sample documentation and laboratory analytical requirements, and equipment calibration and decontamination procedures.

Prior to beginning any field operations, all data-gathering instruments must be properly calibrated using standard calibration media. In addition, all sampling equipment must be decontaminated following the procedures outlined in Chapter 4. Instrument calibration and equipment decontamination is necessary to assure the accuracy and integrity of the data being collected, and to prevent the spread of contamination from the site. After a piece of sampling equipment has been decontaminated it should be carefully wrapped in aluminum foil, with the shiny side of the foil facing outward. This will prevent the equipment from being contaminated prior to the time of sample collection.

A contract with an analytical company should be signed and in place prior to beginning sample collection. This laboratory should provide the appropriate sample bottles, sample preservatives, and information on analytical holding times. Most laboratories will provide sample bottle labels and chain-of-custody seals for sample bottles upon request.

It is recommended that field sampling be delayed until all of the above items have been thoroughly considered. The intrusive sampling methods which are addressed in the following sections include soil-gas surveying, pipe surveying, soil sampling, sediment sampling, surface water sampling, groundwater sampling, and drum sampling.

Soil-Gas Surveying

Soil-gas surveying is one investigative technique that is commonly used to assist remedial investigations where the contaminants of concern include volatile organic compounds. This method is relatively inexpensive when compared to the cost of obtaining similar data through the collection and chemical analysis of individual soil, sediment, and water samples.

Soil-gas in the form of methane occurs naturally in soils as a result of the decomposition of organic material. The concentration of this gas is dependent on a number of environmental factors such as the soil's moisture content, buffer capacity, pH, nitrogen and phosphorus content, and the temperature of the surrounding environment. The optimum conditions for natural soil-gas generation of methane are soils with high moisture content, neutral pH, high nitrogen and phosphorus content, and moderate temperatures (EPA 1989).

In addition to methane gas, it is not uncommon for deep pockets of naturally occurring hydrocarbons to provide trace or, in some cases, stronger detections of various volatile organic compounds at the ground surface. In order to avoid misinterpreting these detections as contamination, it is necessary to collect background soil-gas samples. Background locations should be selected from an undisturbed area upgradient from the study area, and should represent the same soil formation. When the background soil-gas results are subtracted from the study site results, contamination will be revealed. Great care should be taken in selecting the background locations since, depending upon the stratigraphy, soil-gas can migrate upgradient.

The soil-gas technique is most effective in mapping low molecular weight, halogenated solvent compounds and petroleum hydrocarbons possessing high vapor pressures and low aqueous solubilities. These compounds readily partition out of the groundwater and into the soil-gas as a result of their high gas/liquid partitioning coefficients. Once in the soil-gas, volatile organic compounds diffuse vertically and horizontally through the soil to the ground surface, where they dissipate into the atmosphere. Since the contaminants in the ground act as a source, and the atmosphere above the ground surface acts as a sink, a concentration gradient typically develops between the two.

The most effective soil-gas sampling procedures include the Field Screening Method, Microseeps, Mobile Gas Chromatograph, and Petrex Methods. Each of these methods has its own advantages and disadvantages.

The Field Screening Method utilizes a portable flame ionization or photoionization detector in combination with a hand-driven steel sampling rod. Using this method, the investigator is able to sample the relative concentration of organic soil vapors in the field as deep as several feet below the ground surface. The advantage of this method is that sampling results can be obtained quickly and inexpensively. This method is less sophisticated than other methods because the instrument is not able to differentiate between organic compounds; rather, the instrument readings are a measure of relative equivalents to the gas which the instrument is calibrated to (typically methane or pentane). Since this instrument cannot provide results below the one ppm range, it is most commonly used as a "quick and dirty" method of identifying areas where organic contamination is concentrated.

The Microseeps Method utilizes a sophisticated sampling system which uses a large syringe to purge and sample soil gas from a shallow borehole and then transfers the gas into glass vials at a pressure of 10- to 12-psi. Once acquired, the sample can either be analyzed in the field using an onsite laboratory, or sent to the Microseeps laboratory for analysis. Either option provides "laboratory quality" data. The primary advantages of this method are that it is more sophisticated and accurate than the Field Screening Method, since it is able to differentiate between specific organic compounds and can provide results in the parts-per-billion (ppb) range for some compounds. When utilizing the onsite laboratory option, the analytical instrumentation is set up in a facility near the site. The samples are brought to the laboratory, rather than bringing the laboratory to the samples, as with the Mobile Gas Chromatograph Method. The disadvantages of sending the samples to the laboratory as opposed to using the onsite laboratory is a waiting period of several days before analytical results are available. However, when samples are sent to the laboratory for analysis, data for a complete suite of EPA Method 601/602 (32 compounds) can be obtained.

The Mobile Gas Chromatograph Method utilizes a laboratory quality gas chromatograph which is installed in a van or truck. With this method, sampling rods are hydraulically pushed into the ground to collect soil-gas samples. This method has the advantage of providing preliminary analytical results within an hour or so after sampling, with detection limits in the parts-per-billion range. This method also provides the flexibility of collecting samples at intervals as deep as 50 feet in most soils, and allows the collection of samples from more than one interval in the same hole. The primary disadvantage of this method is that sampling is restricted to areas that can be assessed by the mobile laboratory; the method tends to be more expensive than other soil gas methods; and the investigator must select which analytes are to be screened for prior to analyzing the sample so that the instrument can be properly calibrated.

The Petrex Sampling Tube Method utilizes glass sampling tubes which contain two ferromagnetic filaments that are affixed with an activated carbon adsorbent. These tubes are commonly installed one or more feet below the ground surface and left for a period of time. When the tubes are retrieved, they are shipped to a laboratory where the filaments are analyzed using a mass spectrometer. The advantage of the Petrex Method is that a mass spectrometer is used to analyze the sample as opposed to a gas chromatograph, which assures the accurate identification of most organic compounds. The Petrex Method provides the most benefit when investigating sites where little is known about the chemicals present, since the mass spectrometer does not need to be precalibrated for the contaminants of concern, as does the gas chromatograph method. This method also works very effectively in environments where contaminant levels are very low since the sampling tubes can be left in the ground as long as needed to obtain positive results. The disadvantage of this method is that the results are not quantitative, but rather reflect relative variations in soil-gas content and concentrations. A second disadvantage is that it commonly takes several weeks to obtain preliminary analytical results.

For the Microseeps and Petrex Methods, sampling points are commonly arranged in a grid pattern so as to facilitate the contouring of the results. The grid spacing selected should be wide enough to be cost-effective, and yet tight enough to provide definitive results. It is not uncommon to use more than one grid spacing at a site, where a tighter grid is used around areas which are suspected to be sources of contamination. It is also not uncommon to have more than one sampling grid at a site, where one grid is positioned over the suspected source area, and a second grid is positioned over a downgradient ponding area where surface water pools before being absorbed by the soil. Sampling points are also often positioned along drainages leaving the site to identify pockets of contamination (Figure 3.1).

For the Field Screening and Mobile Gas Chromatograph Methods, where sampling results are available shortly after sampling, sampling should begin near the suspected source(s) of contamination and proceed radially outward until the boundaries of the soil-gas plume have been defined (Figure 3.2).

If soil-gas results are to be contoured, it is essential that all of the samples be collected from the same depth interval, and be obtained using the same sampling method. When using the Mobile Gas Chromatograph Method, it is not uncommon to collect soil-gas samples from two depths at each sampling location. In this case, two separate contour maps can be generated, one from each interval.

Soil-gas surveying results are not sufficient by themselves to characterize a site since the results do not reflect the concentration of contaminants in the soil or groundwater. Rather, the results from the soil-gas survey are used to strategically position soil and groundwater sampling locations, where samples can be collected for confirmation laboratory analysis.

Field Screening Method

The Field Screening Method utilizes a portable ionization detector, such as an OVA or HNu meter, along with a hand-driven stainless steel sampling rod to measure the concentration of organic vapors 5-ft or less below the ground surface. The sampling tool is composed of a slide-hammer, extension rod and

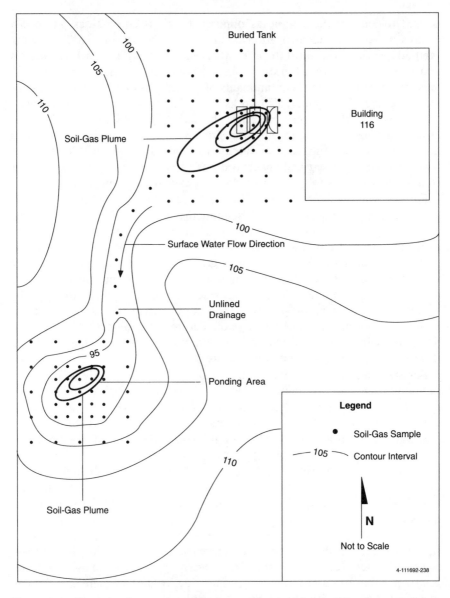

Figure 3.1. Example of a common soil-gas sampling grid for the Microseeps or Petrex Methods.

probe with vapor holes for gas entry, a removable inner liner rod which prevents soil intrusion into the probe, and an extension rod (Figure 3.3). The screening instrument used for this method is less sophisticated than other soil-gas methods since it provides only a measure of the relative equivalents to the gas to which the instrument is calibrated (typically methane or pentane).

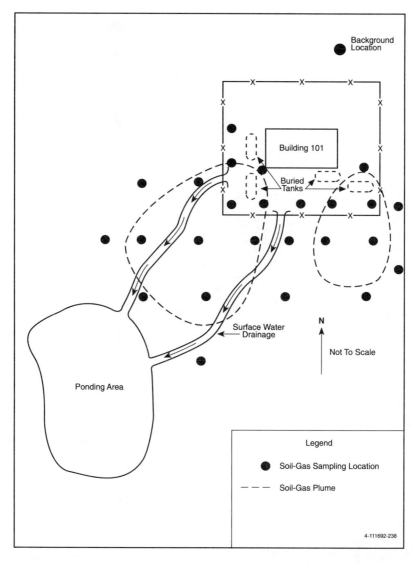

Figure 3.2. Example of a common soil-gas sampling effort using the Field Screening or Mobile Gas Chromatograph Methods.

The advantage of this method is that sampling results can be obtained quickly and inexpensively. Since this technique cannot provide results below the one part-per-million range, it is most commonly used as a "quick and dirty" method of identifying areas where organic contamination is concentrated. If the objective of the soil-gas survey is to precisely define the boundaries of an organic contaminant plume, the Microseeps, Mobile Gas Chromatograph, or Petrex Methods are more appropriate (see pages 63, 66, and 69).

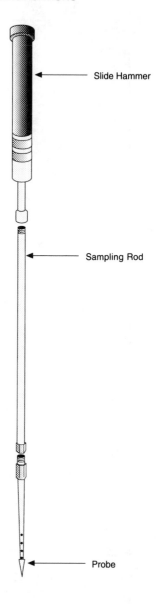

Slide Hammer

Sampling Rod

Probe

Figure 3.3.   Sampler rod used to collect
soil-gas samples for the Field
Screening Method.

For most sampling programs, four people are sufficient for this sampling procedure. Two are needed for inserting and removing the sampling rod, collecting the sample, and documenting the results. A third is needed for health and safety and quality control, and a fourth is needed for managing waste drums and equipment decontamination.

After laying out a sampling grid, the following equipment list and procedure is used to collect soil-gas samples:

1. slide-hammer and soil-gas probe
2. stainless-steel sampling rod and liner
3. portable ionization detector
4. instrument calibration gas
5. sample logbook
6. health and safety screening equipment
7. health and safety clothing
8. DOT approved 55-gal waste drum
9. plastic sheeting

Sampling Procedure

1. Prior to sampling, read the introduction to Intrusive Sampling Methods (page 55) to confirm that all necessary preparatory work has been completed. This preparatory work includes: obtaining utility and property access agreements; defining health and safety, decontamination and waste disposal requirements; and calibrating all health and safety and sampling equipment.
2. After suiting up into the appropriate level of protection, cut a hole in the plastic sheeting several inches in diameter, then lay it over the location to be sampled.
3. Using the slide hammer, beat the sampling rod to the desired sampling depth.
4. Remove the liner from the sampling rod and quickly insert the probe of the ionization detector. Leave the probe in the sampling rod for one minute and record the highest reading.
5. Remove the sampling rod from the ground by either rocking it back and forth several times prior to lifting upward, or by using the slide hammer to reverse beat the rod up out of the hole.
6. Very little, if any, waste material is generated from this sampling procedure; however, whatever waste is generated should be containerized in a DOT approved 55-gal drum. Prior to leaving the site, the waste drum should be sealed, labeled, and handled appropriately (see Chapter 7).
7. At least one point on the sampling grid should be surveyed in by a professional surveyor to provide an exact sampling location. The remainder of the sampling grid can be tied into this one surveyed location.

Microseeps Method

The Microseeps sampling procedure uses a slide hammer, 0.5-in. diameter solid steel drive rod, probe sampler, and glass vials to collect soil-gas samples. The probe sampler consists of a hollow stainless steel rod approximately 4-ft in length through which a 0.25-in. stainless steel tubing has been machine-fitted (Figure 3.4). The bottom of the sampler contains an 8-in. perforated probe tip which allows soil-gas to enter the sampler. Above the probe tip is a glass wool

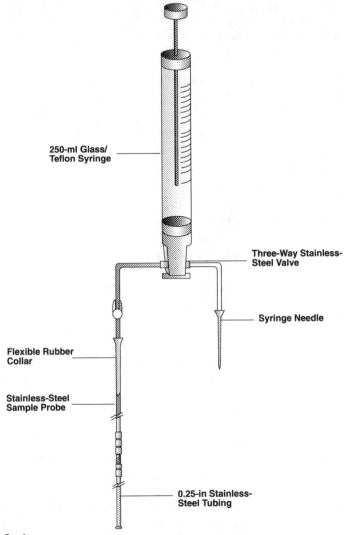

**250-ml Glass/
Teflon Syringe**

**Three-Way Stainless-
Steel Valve**

Syringe Needle

**Flexible Rubber
Collar**

**Stainless-Steel
Sample Probe**

**0.25-in Stainless-
Steel Tubing**

Not to Scale

Figure 3.4.   Sampling tool used to collect soil-gas samples using the Microseeps Method.

filter which prevents soil particles or dust from entering the sample tube above. The upper end of the probe contains a three-way valve to a 250 mL glass/Teflon magnum syringe. The sampling probe is also fitted with a flexible collar to aid in providing an air-tight surface seal. The glass sample vials are fitted with Teflon-faced butyl rubber septa (Wyatt 1992).

The advantage of the Microseeps Method over the Field Screening Method (page 59) is that the sample is analyzed by a gas chromatograph in an onsite or Microseeps laboratory, as opposed to a field ionization detector. This method can therefore identify a specific analytical compound with some detection limits in the parts-per-billion (ppb) range. The Microseeps Method is also much more effective in defining the boundaries of a contaminant plume than the Field Screening Method, which simply helps to define areas where contaminants are concentrated. Although the Mobile Gas Chromatograph Method (page 66) has the advantage of drawing the soil-gas sample directly into the analytical equipment as opposed to a glass vial, the Microseeps Method is typically more cost-effective and is not limited to sampling only areas that can be accessed by a vehicle. The Petrex Method (page 69) provides the advantage of analyzing soil-gas samples using mass spectrometry; however, the analytical results are not quantitative. The Microseeps Method can also provide analytical data in a shorter time than the Petrex Method, since the Petrex Tubes must be left in the ground for as much as one to four weeks.

For most sampling programs, four people are sufficient for this sampling procedure. Two are needed for inserting and removing the sampling rod, collecting the sample, and documenting the results. A third is needed for health and safety and quality control, and a fourth is needed for managing waste drums and equipment decontamination.

After laying out a sampling grid, the following equipment list and procedure is used to collect soil-gas samples:

1. Microseeps probe sampler
2. slide hammer
3. 0.5-in. stainless steel drive rod
4. glass sample vials
5. sample labels
6. cooler packed with Blue Ice®
7. trip blank and coolant blank
8. sample logbook
9. health and safety screening equipment
10. health and safety clothing
11. DOT-approved 55-gal waste drum
12. plastic sheeting
13. sampling table

Sampling Procedure

1. Prior to sampling, read the introduction to Intrusive Sampling Methods (page 55) to confirm that all necessary preparatory work has been completed. This preparatory work includes: obtaining utility and property access agreements; defining health and safety, decontamination and waste disposal requirements; and calibrating all health and safety and sampling equipment.
2. After suiting up to the appropriate level of personal protection, cut a hole in the plastic sheeting several inches in diameter, then lay it over the location to be sampled.

3. Prior to beginning the sampling procedure, the probe sampler must first be purged by drawing ambient air into the syringe through the needle port by lifting up on the syringe piston. The syringe needle should not be screwed into the needle port for the procedure. The three-way valve is then turned to allow the purge air to be ejected from the syringe into the probe sampler by depressing the syringe piston. This procedure should be repeated five times before using the probe to collect a sample.
4. Using the slide hammer, beat the drive rod to the desired sampling depth.
5. Remove the drive rod from the hole and quickly insert the probe sampler. See that the rubber collar is pushed deep enough into the hole to provide an air tight seal.
6. Draw 10-mL of air from the probe into the syringe and expel it through the syringe needle port. This is performed to remove atmospheric air from the probe sampler, and to fill the probe volume with an undiluted soil-gas sample.
7. Screw a syringe needle onto the needle port.
8. Collect a soil-gas sample through the probe sampler by slowly raising the syringe piston to the desired sample volume. Turn the three-way valve and inject the sample through the syringe needle into a sample vial. The sample volume to be collected should be 1.5 to 2.0 times the volume of the sample vial into which the soil-gas sample is compressed. Since this results in a positive pressure inside the sample vial, any leakage of soil-gas through the septum will be out of the vial, resulting in no dilution or contamination of the soil-gas sample.
9. Remove the probe sampler from the ground.
10. Very little, if any, waste material is generated from this sampling procedure; however, whatever waste is generated should be containerized in a DOT-approved 55-gal drum. Prior to leaving the site, the waste drum should be sealed, labeled, and handled appropriately (see Chapter 7).
11. At least one point on the sampling grid should be surveyed in by a professional surveyor to provide an exact sampling location. The remainder of the sampling grid can be tied into this one surveyed location.

Mobile Gas Chromatograph Method

With the Mobile Gas Chromatograph Method, soil-gas samples are analyzed in the field by a mobile laboratory immediately following sample collection. With this method, preliminary analytical results from a soil-gas sample are commonly available within an hour after sampling. The number of analytes being screened and the sampling depth directly affect the sampling time; however, for planning purposes one can expect to collect approximately 8 to 12 samples per day using this method.

A mobile laboratory is commonly equipped with one or more laboratory grade gas chromatographs, a temperature-programmable oven, and various types of instrument detectors such as electron capture, flame ionization, photoionization, and thermoconductivity. It also contains a computer to record the instrument output, and a generator to make a self-contained operation. Vendors who provide this service provide a chemist to operate the analytical equipment, and an environmental technician to collect the samples.

The analytical equipment must be calibrated daily for each of the analytes to be screened for. The calibration is performed each morning prior to the collection of the first sample. The time required for calibration is dependent on the number of analytes being screened for. In general, the calibration procedure takes between one and two hours.

Soil-gas samples are collected by pushing a small diameter steel rod into the ground to the desired sampling depth. To perform this operation, the mobile laboratory uses a hydraulic press which is mounted on the rear end of the vehicle (Figure 3.5). A small disposable nose cone is placed on the bottom end of the sampling tube prior to pushing it into the ground to prevent soil from entering the tube. For deeper penetration, a conepenetrometer can be used to perform this operation.

Once the desired sampling depth is reached, an air evacuation line is used to purge the tube several times in preparation for sampling. A sample of the air is then extracted and injected into a gas chromatograph. Common laboratory quality assurance procedures, such as running laboratory duplicate and spike samples, are used to ensure the accuracy and precision of the analytical results.

The soil-gas detection limits are a function of the injection volume as well as the detector sensitivity for individual compounds. Generally, the larger the injection size, the greater the sensitivity. However, peaks for compounds of interest must be kept within the linear range of the detector. If any compound has a high concentration, it is necessary to use small injection volumes and, in some cases, dilute the sample to keep it within linear ranges. This may result in decreased detection limits for the other compounds in the analysis.

When using this soil-gas method, sampling should begin at areas known to be, or suspected to be the most contaminated. Samples are then collected radially away from these areas until soil-gas concentrations drop to background levels. Since soil-gas samples can be collected at various depths, this technique can provide useful information regarding the depth of soil contamination.

A step-by-step procedure has not been provided for this method since this operation must be performed by professionals specifically trained in operating the sampling tools and instruments.

The advantage of the Mobile Gas Chromatograph Method over the Field Screening Method (page 59) is that the sample is analyzed by a gas chromatograph, as opposed to an ionization detector in the field. This method can therefore identify specific analytical compounds with some detection limits in the ppb range. The Mobile Gas Chromatograph Method is also much more effective in defining the boundaries of a contaminant plume than the Field Screening Method, which simply helps to define areas where contaminants are concentrated. The Mobile Gas Chromatograph Method has the advantage of

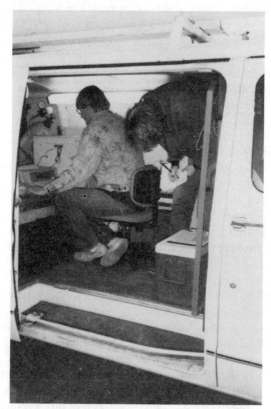

Figure 3.5.   Sampling tool and instrument used to collect and analyze soil-gas samples using the Mobile Gas Chromatograph Method.

providing analytical results to investigators while they are still in the field. However, this method is limited to areas which can be accessed by the sampling vehicle. The Petrex method provides the advantage of analyzing soil-gas samples using mass spectrometry; however, the analytical results are not quantitative, as the Mobile Gas Chromatograph samples are.

Petrex Sample Tube Method

This method utilizes Petrex sampling tubes installed below the ground surface to collect soil-gas samples. The Petrex tube is comprised of a large glass test tube which contains two ferromagnetic wire filaments coated with an activated carbon adsorbent. Each tube has an airtight cap to seal the tube, and a retrieval wire for removing the sampler from the ground (Byrnes 1990).

Petrex tubes are most commonly installed between 6-in. and 3-ft below the ground surface in a grid pattern over a study area (Figure 3.6). Whatever sampling depth is chosen, it is critical that all the tubes from one sampling grid be set at the same depth so that the analytical results are comparable. The sample tubes are left in the ground anywhere from one day to four weeks before retrieval. The length of the field exposure period depends on the rate at which a significant portion of the survey wires becomes "loaded" (NERI 1992). If sampling tubes are set deeper than 3-ft, a metal pipe can be used to protect the sampler from being contaminated by near-surface contaminants (Figure 3.7).

To assess the loading rate of a particular survey, typically five Petrex tubes are installed onsite strictly for the purpose of time calibration. These time calibration tubes are installed at given survey grid locations where soil-gas concentrations are expected to be the highest. Approximately one day after installation, several of the time calibration tubes are retrieved and shipped overnight to the laboratory for analysis. The remaining time calibration tubes are retrieved for analysis at intervals determined by earlier testing. In most cases, samples can be retrieved on the basis of a single set of time calibration analyses (NERI 1992).

If sample tubes are overexposed to the soil-gas, it is difficult for the mass spectrometer to accurately differentiate between organic compounds. If the tubes are underexposed, an accurate delineation of the contaminant plume cannot be made. To prevent the disturbance of soil-gas equilibrium, time calibration tubes should not be installed any closer than 5 ft from a sample tube.

The second filament in the sample tube can be used as a duplicate sample, where one duplicate sample is commonly run for every ten samples analyzed. These tubes are used for quality control purposes to determine the precision of the analytical results.

The advantage of the Petrex Method over the Field Screening Method (page 59) is that the sample is analyzed by a mass spectrometer, as opposed to an ionization detector in the field. This method can therefore identify specific

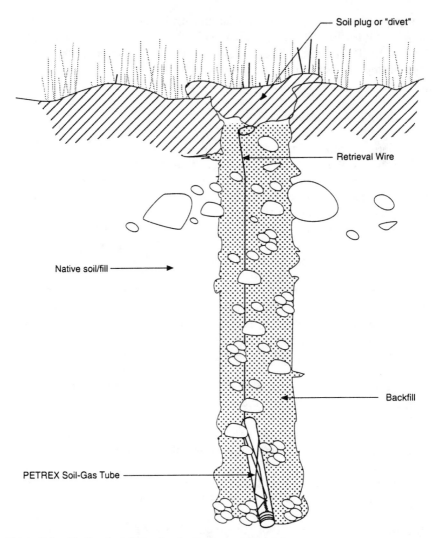

Figure 3.6. Shallow installation for the Petrex sample tube.

analytical compounds. The Petrex Method is also much more effective in defining the boundaries of a contaminant plume than the Field Screening Method, which simply helps to define areas where contaminants are concentrated. The Petrex Method is very effective in characterizing sites with very low concentrations of contaminants since the sampling tubes can be left in the ground as long as needed to obtain positive results. The disadvantages of this method are that it takes longer to obtain analytical results than the other methods, and the results are not quantitative but rather reflect relative variations in soil-gas content and concentrations.

For most sampling programs, four people are sufficient for this sampling

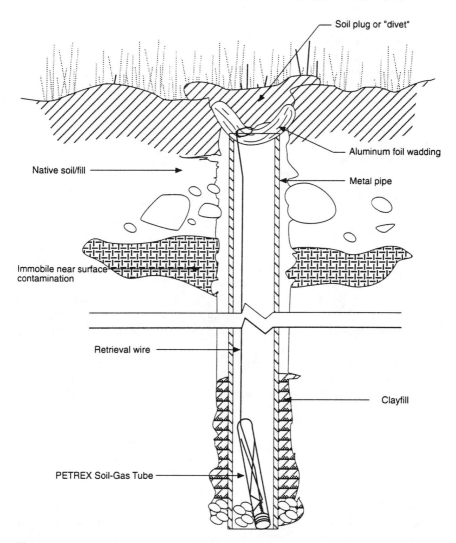

Figure 3.7.   Deep installation for the Petrex sample tube.

procedure. Two are needed for installation and retrieval of sampling tubes, and documentation. A third is needed for health and safety and quality control, and a fourth is needed for managing waste drums and equipment decontamination.

After the sampling grid has been established, the following equipment list and procedure is used to install and retrieve soil-gas tubes:

1. stainless steel hand auger
2. stainless steel spoon
3. stainless steel bowl

4. trowel and chisel
5. stainless steel tongs
6. Petrex soil-gas sampling tube
7. retrieval wire
8. metal pipe (2-in. diameter)
9. sample labels
10. sample logbook
11. cooler packed with Blue Ice®
12. trip blank
13. health and safety screening equipment
14. health and safety clothing
15. plastic sheeting (5' × 5')
16. fluorescent flagging tape

Sampling Procedure

1. Prior to sampling, read the introduction to Intrusive Sampling Methods (page 55) to confirm that all necessary preparatory work has been completed. This preparatory work includes: obtaining utility and property access agreements; defining health and safety, decontamination and waste disposal requirements; and calibrating all health and safety and sampling equipment.
2. After suiting up to the appropriate level of protection, cut a hole in the plastic sheeting several inches in diameter, then lay it over the point to be sampled. The purpose of this sheeting is to prevent the spread of contamination.
3. Using a stainless steel auger, drill a hole to the desired sampling depth. Each time the auger is removed from the hole, use a stainless steel spoon to transfer the cuttings from the auger into a stainless steel bowl.
4. Remove the airtight cap from the Petrex sample tube and lower it to the bottom of the hole using a retrieval wire.
5. Backfill the borehole with the original excavated soil and sod as shown in Figure 3.6. If a metal pipe is used for deeper installation, backfill soil around the outside of the pipe. To prevent soil from entering the inside of the pipe, use a piece of wadded aluminum foil, as shown in Figure 3.7.
6. Record in a sample logbook the time and date that the tube was installed.
7. Attach a piece of fluorescent flagging tape to the end of the wire line. Write the sample and station number on the tape using a permanent ink marker. This flagging should be visible above the ground surface. The sample tube should also be labeled with the sample number.
8. Use a trowel and chisel to expose the backfilled sampler. Using a pair of stainless steel tongs, grab the retrieval wire and remove the sampler.
9. After the sampler is capped, attach a sample label and custody seal, then immediately place it in a Blue Ice®-packed cooler.
10. See Chapter 5 for details on preparing samples for shipment.
11. In order to prevent the spread of contamination, sample holes should be

filled with bentonite grout, and soil cuttings should be contained in a DOT-approved 55-gal drum. Prior to leaving the site, the waste drum should be sealed, labeled, and handled appropriately (see Chapter 7).

12. At least one point on the sampling grid should be surveyed in by a professional surveyor to provide an exact sampling location. The remainder of the sampling grid can be tied into this one surveyed location.

Pipe Surveying

The following section provides the reader with information regarding pipe surveying tools which are available to assist a remedial investigation study. The primary objective behind performing a pipe survey is to determine if contaminated soil or other material is present in the pipe, determine if there are any cracks in the pipe where the contamination could leak into the surrounding formation, and determine where the pipe ultimately discharges. This information is used to help focus a remedial investigation around areas that are more likely to be contaminated. For example, if cracks are identified in a pipe, soil-gas surveying and/or soil sampling can be focused around these areas as opposed to characterizing the soil surrounding the entire length of the pipe. If contaminated soil is found within a pipe, special tools are available which can clear this material from the pipe. Currently, pipe surveying technology is undergoing rapid advancement due to its practical applicability, and its relative cost-effectiveness.

Surveying tools are advanced through a pipe using either a self-propelled crawler unit or fiberglass push cable. The crawler unit is currently commercially available in sizes small enough to investigate 6-in. diameter pipes. Smaller diameter pipes can be investigated by advancing tools using a fiberglass push cable. Some of the investigative tools which are currently available for pipe surveying include:

- high resolution video cameras
- organic vapor analyzers
- combustible gas indicators
- alpha, beta, and gamma radiation detectors, and
- temperature and pH meters.

Pipe crawler units are commonly designed to meet site-specific requirements and conditions, such as:

- pipe diameters
- pipe bend angles
- pipe surface texture
- surveying distances
- analytical instrumentation, and
- site radiation or chemical contamination levels.

Depending on the pipe geometry, it is often necessary to build the crawler unit in modules, which are linked together in a train-like configuration. When a modular system is used, a separate module can be used for the motor, screening instruments, and video camera(s) (Figure 3.8).

Wheels or tracks are used to propel the crawler units. Wheeled units use four or six wheels which are narrow, and chamfered to the approximate radius of the pipe. These wheels combined with a Direct Current (DC)-powered motor provide traction to negotiate most sludge and mud-like environments (Figure 3.9).

Tracked systems can be advantageous for short distance applications (< 100-ft) in large diameter pipes or ducts, due to their ability to turn on their own center. This feature can be very useful when negotiating complex runs, particularly when there is a significant amount of debris. With this one exception, wheeled systems outperform tracked systems in most other applications.

There are some very sophisticated devices that have been developed for characterizing pipes that utilize what is known as inchworm or spider technology. These types of devices can negotiate small diameter piping or tubing

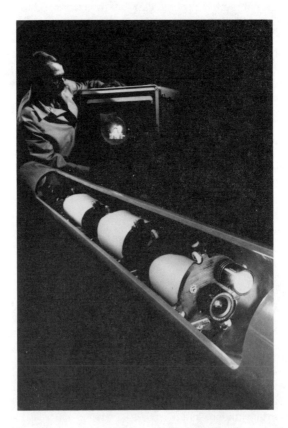

Figure 3.8. Modular pipe surveying instrument linked together in a train-like configuration.

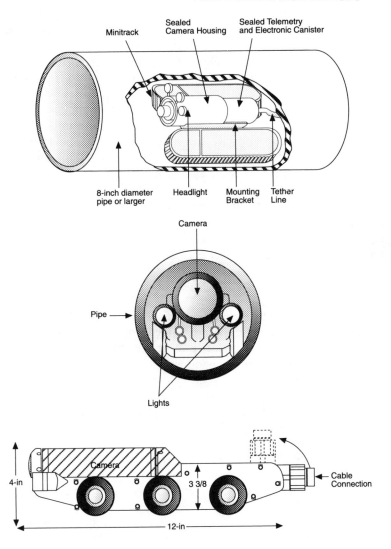

Figure 3.9.   Example of a wheel and track crawling unit (RSI Research Ltd., Sidney, British Columbia, Canada).

systems. However, these devices are very slow, and are more prone to break-down than the simpler wheel or track units.

The distance which a system can travel is primarily dependent on the weight of the tractor package, and the size and weight of the telemetry cable. As a general rule, a tractor can only pull its own weight in cable and associated drag. Onboard multiplexing or using fiber optic telemetry transmission cable can help to reduce the drag. However, fiber optic cable is very fragile and therefore is not recommended since piping interrogation activities are commonly performed in rugged environments. Some wireless telemetry controlled systems have also been developed; however, they tend to be unreliable. When using wireless systems in metal pipes there is the potential for confusing the transmission signal receiver, and thus there is the risk of losing a system in the pipe.

Another approach that can be used for long distance application (> 100-ft) is powering the crawler and the sensor instruments with onboard batteries. With this approach, the cables are only needed to provide trickle charge to the crawler package and to transmit information. Maximum horizontal distances achieved at this time exceed 1,500-ft.

Some accessories which can be designed modularly, and that can be utilized to expand the capability of the crawler unit include:

- pan-and-tilt for vision and/or directional sensors
- laser distance and scaling instrumentation
- pipe clearing tooling, and
- inclinometry for measuring settling or slope.

A new technology that is being investigated uses chemically sensitive microchips to identify and measure specific elements or chemical species within the pipe system. Two other technologies in development at this time include a sensor package to detect contaminants that have leaked from a breached piping system, and an in situ pipe repair system.

If a specific detector requires the pipe to be free of debris prior to performing the survey, a squeegee-like tool can be attached to a crawler unit (Figures 3.10 and 3.11). Tethers are attached to all devices to allow retrievability in the event the drive unit fails.

The Manual Push Method is generally required for pipes smaller than 6-in. in diameter, and utilizes 1/8-in. to 5/8-in. solid flexible polyethylene rods to push and retrieve sampling tools (Figure 3.12). At this time, as far as the author is aware, the longest push accomplished in mockup is approximately 700-ft, and approximately 350-ft in actual field application. The same investigative tools available for the pipe crawler are also available for the Push Method. In order to keep the sampling instrument or camera properly positioned in the pipe, instrument-bearing skids are used. For larger diameter pipes (3- to 6-in.), wheeled "dogbone" shaped skids are most effective, while skids with runners are used in smaller diameter pipes. In an effort to maximize distance, small tri-wheeled devices can be attached to the push-rod every several feet to reduce the friction of the cable rubbing against the pipe.

The manual push approach is very cost-effective at approximately $2.00/ft for the push cable. If short distances (400-ft) are the intent, then the Push Method can be effectively utilized in pipes up to 10-in. in diameter. This approach is less effective for larger diameter piping since the push rod can bunch up more easily and thus reduce the acquired distances.

Information regarding the location or position of a sampling tool in a pipe

Figure 3.10. Squeegee-like pipe cleaning tool.

Figure 3.11. Squeegee-like pipe cleaning tool removing sediment from a pipe.

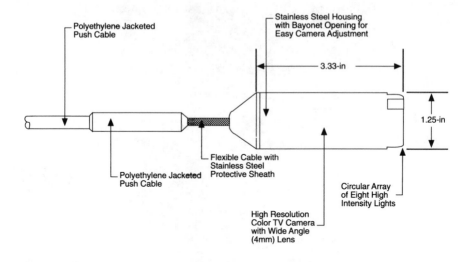

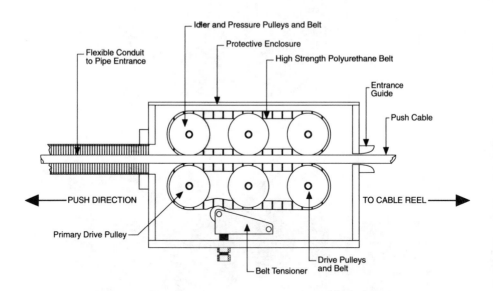

Figure 3.12. Small-diameter television camera and push cable mechanism (RJ Electronics, Turner, Oregon).

can be acquired by means of a footage encoder, pressure wheels and transducers, or inclinometers, which are all commercially available. Fiber optic encoders have also been used with some success, although the setup of such devices is more complex.

Software systems are currently being developed that can provide precise system positioning information from a photogrammetry-based program. The only requirement for this method is a physical reference point within the area of interest. This method can also be used to navigate equipment within an area remotely, and may replace more complex methods. Information can be displayed and recorded on a cathode ray tube (CRT) or digital display indicating distance, pipe identification number, date, and time.

There is instrumentation now available that accepts real-time voltage or current inputs, translates them into reportable values (i.e., RAD/hr), and displays the results real-time on the video screen. This device has a high/low level alarming capability and will accept analog, log, nonlinear or quadrature inputs. This instrument is particularly valuable for providing a single source documentation package of raw data and acquired information.

Imaging systems are available in inexpensive models, which can easily be damaged in radioactive environments, or radiation-resistant models which are built to withstand higher levels of radiation. The two basic types of cameras available include the tube and the Charged Coupled Device (CCD). The tube design cameras are limited by high lighting requirements. The CCD systems are typically much less radiation-resistant but have the advantage of low lighting requirements and can be utilized in the Black-White/Chroma format. Radiation resistance testing performed on some CCDs has shown a significant difference in radiation resistance from one model to another. Overall, they are all able to last to approximately the low to mid-10^4 RAD range for accumulated dose. For short-term exposure, the limits vary from 20 RAD/hr to 200 RAD/hr. As the exposure levels increase, the video screen begins "sparkling." High level exposures can completely white-out the image. This effect goes away once the camera is removed from the radiation field.

Another relatively new technology utilizes charge injection device (CID), which manufacturers claim has significant radiation resistance. The accumulated dose threshold for this camera is not significantly higher than the CCD at 10^5 RAD.

Tube cameras have significant radiation resistance capability. Short-term dose effects are not seen until approximately 10^5 RAD, while the long-term accumulation threshold is in the 10^8 RAD range. The tradeoff occurs in the high light requirements and long length of the color system. The black-and-white system has the problems associated with black-and-white images. The latest in radiation-resistant color CCD cameras is a lead shielded, optics-line-of-sight-removed system that reportedly provides even longer life and functionality in radiation environments than the tubed cameras for even less cost. These systems have high resolution and zoom capability and can employ CCD front ends that can provide clear imaging in the < 1 Lux illumination range. The disadvantage of these systems is the bulky package and weight. Diameters of the camera are approximately 3.5-in. and the lead-shielded unit weighs approximately 30 lb.

One of the most sophisticated large pipe surveying tools is the Hanford Information Gathering Instrument (HIGI) developed in 1991 by the Engineering Surveillance and Testing group (Figure 3.13). This system was designed and developed to assess two large diameter pipelines at the Hanford Site that carried effluent water from a reactor to cooling ponds. The system was used to identify areas along the pipe where leaks may have occurred in the past. This information was used to focus a soil sampling effort. HIGI contains the following features:

- closed-circuit television camera which can pan and tilt
- three banks of spotlights
- two four-wheel-drive miniature tractor/crawlers
- 1,500-ft of multiplex signal/receiver cable with linear encoder
- 1,500-ft of aircraft cable umbilical cord
- radiation monitor and a combustible gas/oxygen meter
- size: 2.5-ft high, 2.0-ft wide, and 3.0-ft long, and
- weight: 350 lb.

These investigative tools are particularly useful at old chemical or industrial plants where there are no records showing where historical pipes and drainages

60-inch Application

Figure 3.13. Hanford Information Gathering Instrument (HIGI).

flow, or what the pipes and drains were used for. These tools help reduce health and safety risks to workers by investigating areas where there is a potential lack of oxygen (confined space), or other related hazards. Examples of other investigative tools and setups are shown in Figures 3.14 and 3.15.

Soil Sampling

The following section provides the reader with guidance on selecting shallow and deep soil sampling methods for remedial investigation studies. The criteria used in selecting the most appropriate method include the analyses to be performed on the sample, the type of sample being collected (grab or composite), and the sampling depth. SOPs have been provided for each of the methods to facilitate implementation.

One objective of an initial site characterization study should be to determine if contamination is present in the soil surrounding each of the potential sources of contamination. Some of the more common causes of soil contamination problems include chemical spills, leaking drums or storage tanks, and improper waste disposal practices. If contaminants are identified in the soil, it is important to define both the vertical and horizontal extent of contamination.

If the contaminants of concern are volatile organics, it is typically most cost-effective to define the sources and horizontal extent of contamination using soil-gas surveying combined with confirmation soil sampling. Some soil-gas sampling techniques can also provide assistance in defining the depth of contamination (see Soil-Gas Surveying section, page 57). If the contaminants of concern are nonvolatile, other Secondary Characterization Tools such as magnetics or ground penetrating radar may be used to help focus a soil characterization effort around those areas that have the greatest probability of being contaminated. If a systematic or random soil sampling strategy is selected, the grid spacing selected should be based on probability requirements as discussed on page 33.

In addition to defining or confirming the boundaries of a soil contamination plume, soil samples are also collected in the field for geotechnical testing. These types of analyses are important when evaluating contaminant migration rates, and the feasibility of various remedial alternatives. The following sections present the most effective shallow and deep soil sampling methods, along with detail procedures on how to use them.

Shallow Soil Sampling

Soil samples collected from a depth of 5-ft or less are generally referred to as "shallow." The most effective shallow soil sampling methods include the Scoop, Hand Auger, Slide-Hammer, Open-Tube, Split-Tube or Solid-Tube, and Thin-Walled Tube. When preparing a sampling program, considerable

Figure 3.14. Examples of other pipe surveying tools.

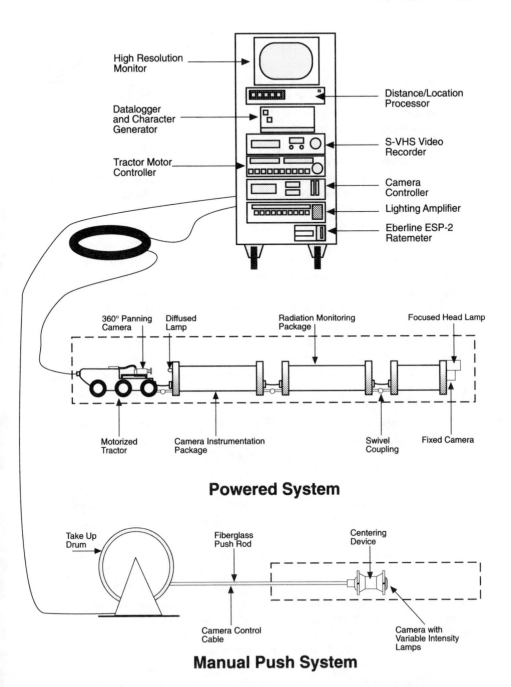

High Resolution
Monitor

Distance/Location
Processor

Datalogger
and Character
Generator

S-VHS Video
Recorder

Tractor Motor
Controller

Camera
Controller

Lighting Amplifier

Eberline ESP-2
Ratemeter

360° Panning
Camera

Diffused
Lamp

Radiation Monitoring
Package

Focused Head Lamp

Motorized
Tractor

Camera Instrumentation
Package

Swivel
Coupling

Fixed Camera

Powered System

Take Up
Drum

Fiberglass
Push Rod

Centering
Device

Camera Control
Cable

Camera with
Variable Intensity
Lamps

Manual Push System

Figure 3.15.   Figure showing cable connections from surveying instruments to control and
monitor instrumentation.

thought should go into selecting appropriate sampling methods, since the selected method can influence the analytical results. For example, if the contaminants of concern at a site are volatile organics, it would not be good practice to collect samples using the Hand Auger Method, since this method churns up the soil, which facilitates volatilization. The Slide-Hammer, or Split-Tube or Solid-Tube Sampler would be a more appropriate selection, since both of these tools can be used to remove a compacted, but undisturbed core of soil.

The Hand Auger is most effectively used to collect composite soil samples from sites where the contaminants of concern do not include volatile organics. When using this tool, samples should not be composited over intervals greater than several feet, since compositing larger intervals tends to dilute the sample beyond the point of providing useful data.

The Slide-Hammer and Split-Tube or Solid-Tube samplers can be used to collect either grab or composite samples. When collecting a grab sample, these tools are commonly lined with sample sleeves which can be quickly capped after sample collection. When composite samples are collected, the soil from the interval(s), or locations to be composited, is transferred into a stainless steel bowl and homogenized with a stainless steel spoon prior to filling sample jars.

The Scoop sampler can be used to collect a grab or composite samples of surface soil. To collect a grab sample, surface soil is scooped directly into a sample jar. A composite sample can be collected by scooping surface soil from the locations to be composited into a stainless steel bowl, and homogenized prior to filling sample jars.

When shallow soil samples are needed for lithology description only, the Open-Tube sampler is an effective tool. This tool has a sampling tube of small enough diameter that it can easily be advanced several feet into the ground to provide a small diameter soil core. Since the sampling tube is open on one side, the soil lithology can be described without removing the sample from the tube. When soil samples are needed for geotechnical analysis, the Thin-Walled Tube sampler hydraulically pushed into the ground is the preferred sampling method.

Table 3.1 summarizes the effectiveness of each of the six recommended sampling methods. A number "1" in the table indicates that a particular procedure is most effective in collecting samples for a particular laboratory analysis, sample type, or sampling depth. A number "2" indicates that the procedure is acceptable, but less effective, while an empty cell indicates that the procedure is not recommended. For example, Table 3.1 indicates that the Slide-Hammer and Split/Solid-Tube Methods are most effective in collecting soil samples for volatile organic analysis. The Scoop Method is considered acceptable when collecting samples for this analysis, while the Hand Auger, Open-Tube, and Thin-Walled Tube Methods are not recommended. The following sections provide further details and SOPs for each of the recommended soil sampling methods.

Table 3.1. Evaluation Table for Shallow Soil Sampling Methods

	Laboratory Analyses								Sample Type			Sampling Depth		Lithology Description
	Volatiles	Semi-Volatiles	Primary Metals	Pesticides	PCBs	TPH	Radionuclides	Geotechnical	Grab	Composite (Vertical)	Composite (Areal)	Surface (0-0.5 ft.)	Shallow (0.5-5.0 ft.)	
Scoop	2	1	1	1	1	1	1		1		1	1		1
Hand Auger		1	1	1	1	1	1			1	2	1	1	1
Slide-Hammer	1	1	1	1	1	1	1		1	1	2	1	1	2
Open-Tube									1			1	1	1
Split-Tube/Solid-Tube	1/1	1/1	1/1	1/1	1/1	1/1	1/1		1/1	1/2	2/2		1	1/2
Thin-Walled Tube								1	1				1	

1 = Preferred Method
2 = Acceptable Method
Empty Cell = Method Is Not Recommended

Scoop Method. The scoop is a hand-held sampling tool which is effective in collecting samples of the top 0.5-ft of soil (Figure 3.16). This method is commonly used to collect samples of discolored soil observed at the ground surface, or to collect samples from areas where, for some reason, deeper sampling is not possible. Grab or composite samples can be collected with this method by either spooning soil from one location directly into a sample jar, or by

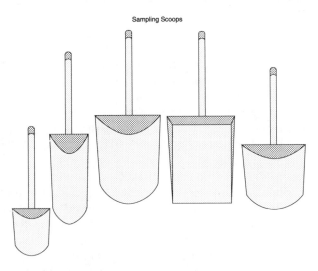

Sampling Scoops

Figure 3.16.   Soil sampling scoops of various sizes and shapes.

compositing soil in a stainless steel bowl from more than one location prior to filling a sample jar.

For most sampling programs, four people are sufficient for this sampling procedure. Two are needed for sample collection, lithology description, labeling, and documentation, a third is needed for health and safety and quality control, and a fourth is needed for miscellaneous tasks such as waste management, and equipment decontamination.

The following equipment and procedure can be used to collect shallow soil samples for chemical and/or radiological analysis:

1. stainless steel scoop
2. stainless steel bowl
3. stainless steel spoon
4. sample jars
5. sample labels
6. cooler packed with Blue Ice®
7. trip blank and coolant blank
8. sample logbook
9. chain-of-custody forms
10. chain-of-custody seals
11. permanent ink marker
12. health and safety screening instruments
13. health and safety clothing
14. DOT-approved 55-gal waste drum
15. sampling table
16. plastic sheeting
17. plastic waste bags

Sampling Procedure

1. In preparation for sampling, read the introduction to Intrusive Sampling Methods (page 55) to confirm that all necessary preparatory work has been completed, including: obtaining property access agreements; meeting health and safety, decontamination, and waste disposal requirements; and calibrating all health and safety and sampling equipment.
2. Cut a 1-ft diameter hole in the center of the plastic sheeting, and center the hole over the sampling point. The purpose of this sheeting is to help prevent the spread of contamination.
3. Begin collecting the sample by applying a downward pressure on the scoop until the desired sampling depth is reached, then lift. If a grab sample is being collected, transfer the soil from the scoop directly into a sample jar. If a composite sample is being collected, transfer the soil from each location to be composited into a stainless steel compositing bowl and homogenize with a stainless steel spoon prior to filling a sample jar.
4. After the jar is capped, attach a sample label and custody seal to the jar and

immediately place it into a Blue Ice®-packed cooler. Samples to be analyzed for radionuclides do not commonly require cooling.

5. See Chapter 5 for details on preparing sample jars and coolers for sample shipment.

6. Any soil left over from the sampling should be containerized in a DOT-approved 55-gal drum. Prior to leaving the site, all waste drums should be sealed, labeled, and handled appropriately (see Chapter 7).

7. Finally, the coordinates of the sampling point should be surveyed in by a professional surveyor to preserve the exact sampling location.

Hand Auger Method. The Hand Auger is an effective shallow soil sampling tool when the contaminants of concern do not include volatile organics, since the augering motion facilitates volatilization. This tool is composed of a bucket auger, which comes in various shapes and sizes, a shaft, and a T-bar handle (Figure 3.17). Extensions for the shaft are available to allow sampling

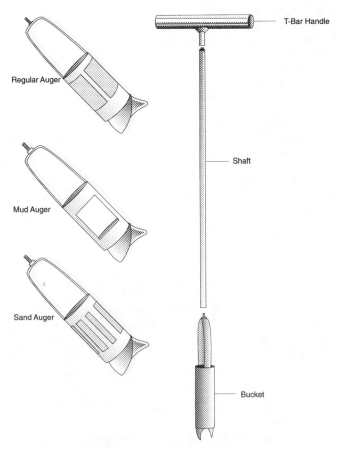

Figure 3.17. Hand Auger used to collect shallow soil samples.

at deeper intervals. However, in most soils this tool is only effective in collecting samples to a depth of 5-ft, since the sample hole typically begins to collapse at this depth.

Since the auger rotation automatically homogenizes the sampling interval, this method is most effective in collecting composite samples. For most characterization studies, samples should not be composited over intervals greater than several feet, since compositing larger intervals tends to dilute the sample beyond the point of providing useful data.

For most sampling programs, four people are sufficient for this sampling procedure. Two are needed for sample collection, lithology description, labeling, and documentation; a third is needed for health and safety and quality control; and a fourth is needed for miscellaneous tasks such as waste management, and equipment decontamination.

The following equipment and procedure can be used to collect shallow soil samples for chemical and/or radiological analysis:

1. stainless steel hand auger
2. stainless steel bowl
3. stainless steel spoon
4. sample jars
5. sample labels
6. cooler packed with Blue Ice®
7. trip blank and coolant blank
8. sample logbook
9. chain-of-custody forms
10. chain-of-custody seals
11. permanent ink marker
12. health and safety screening instruments
13. health and safety clothing
14. DOT approved 55-gal waste drum
15. sampling table
16. plastic sheeting
17. plastic waste bags

Sampling Procedure

1. In preparation for sampling, read the introduction to Intrusive Sampling Methods (page 55) to confirm that all necessary preparatory work has been completed, including: obtaining property access agreements; meeting health and safety, decontamination, and waste disposal requirements; and calibrating all health and safety and sampling equipment.
2. Cut a 1-ft diameter hole in the center of the plastic sheeting, and center the hole over the sampling point. The purpose of this sheeting is to help prevent the spread of contamination.
3. Begin collecting the soil sample by applying a downward pressure while rotating the auger clockwise. When the auger is full of soil it should be

removed from the hole, and the soil transferred into a stainless steel bowl using a stainless steel spoon. Continue sampling in this manner until the bottom of the sampling interval is reached.

4. Composite the soil in the sampling bowl by using the stainless steel spoon to break apart any large chunks of soil, then mix and stir the soil enough to thoroughly homogenize the sample.

5. Transfer soil into a sample jar using the stainless steel spoon.

6. After the jar is capped, attach a sample label and custody seal to the jar and immediately place it into a Blue Ice®-packed cooler. Samples to be analyzed for radionuclides do not commonly require cooling.

7. See Chapter 5 for details on preparing sample jars and coolers for sample shipment.

8. Any soil left over from the sampling should be containerized in a DOT-approved 55-gal drum. Prior to leaving the site, all waste drums should be sealed, labeled, and handled appropriately (see Chapter 7).

9. Finally, the coordinates of the sampling point should be surveyed in by a professional surveyor to preserve the exact sampling location.

Slide Hammer Method. For collecting shallow soil core samples for chemical and/or radiological analysis, the slide hammer coring tool is recommended. This tool is comprised of a stainless steel core barrel, an extension rod, and a slide hammer (Figure 3.18). Most core barrels have an inside diameter of 2- or 2.5-in., and are 1- or 2-ft in length, but can be special-ordered in other sizes. The top of the barrel is threaded so that it can be screwed into an extension rod. The barrel is also constructed so that it can accept stainless steel sample liners which are commonly used to facilitate the removal of soil from the barrel without disturbing the sample. Without sample liners, soil must be extracted from the barrel using a spoon or knife, then transferred into a sample jar. In this procedure, a good portion of the volatile organics can be lost into the ambient air. In contrast, sample liners can be quickly removed from the barrel and sealed with airtight Teflon caps. After labeling the liners, they can be shipped directly to the laboratory for analysis. If samples are being collected for lithology description only, clear plastic liners are available.

Extension rods are available in various lengths to allow sampling at depths greater than the length of the core barrel. These rods are screwed into the core barrel at one end, and into the slide hammer at the other end. A slide hammer is used to beat the core barrel into the ground. The hammer is available in different shapes and weights to accommodate the needs of the sampler.

If a sampling program requires soil core samples to be collected at two-foot intervals, it is recommended that a two-foot core barrel be used to collect the samples, as opposed to collecting two samples using a one-foot barrel. The latter method increases the sampling time, and allows the opportunity for "sluff" soil to be incorporated into the second sample.

For most sampling programs, four people are sufficient for this sampling

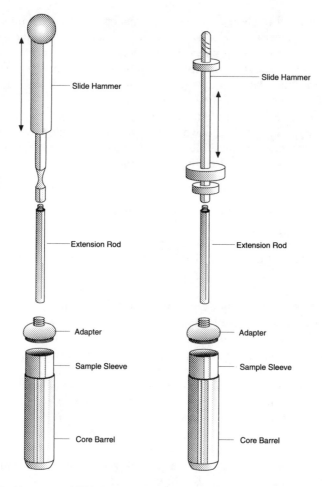

Figure 3.18. Two types of Slide Hammers used to collect shallow soil samples.

procedure. Two are needed for sample collection, lithology description, labeling, and documentation; a third is needed for health and safety and quality control; and a fourth is needed for miscellaneous tasks such as waste management, and equipment decontamination.

The following equipment and procedure can be used to collect shallow soil samples for chemical and/or radiological analysis:

1. slide hammer and extension rods
2. stainless steel sample sleeves
3. Teflon end-caps for stainless steel sleeves
4. stainless steel bowl
5. stainless steel spoon
6. stainless steel knife

7. sample jars
8. sample labels
9. cooler packed with Blue Ice®
10. trip blank and coolant blank
11. sample logbook
12. chain-of-custody forms
13. chain-of-custody seals
14. permanent ink marker
15. health and safety screening instruments
16. health and safety clothing
17. DOT-approved 55-gal waste drum
18. sampling table
19. plastic sheeting
20. aluminum foil
21. plastic waste bags

Sampling Procedure

1. In preparation for sampling, read the introduction to Intrusive Sampling Methods (page 55) to confirm that all necessary preparatory work has been completed, including: obtaining property access agreements; meeting health and safety, decontamination, and waste disposal requirements; and calibrating all health and safety and sampling equipment.
2. Cut a 1-ft diameter hole in the center of the plastic sheeting, and center the hole over the sampling point. The purpose of this sheeting is to help prevent the spread of contamination.
3. If sample sleeves are to be used, unscrew the core-barrel from the slide hammer and load it with decontaminated stainless steel sleeves of the desired length. Avoid touching the inside surface of the core-barrel and sleeves, for this will contaminate the sampler. Screw the core-barrel back onto the slide hammer.
4. Use the slide hammer to beat the core-barrel to the desired depth, and record the blow count in a sample logbook.
5. Remove the core-barrel from the hole by rocking it from side to side several times before lifting or reverse beating the sampler from the hole.
6. To collect a grab sample, unscrew the core barrel from the sampler and slide the sample sleeves out onto a piece of aluminum foil. Using a stainless steel knife, separate the sample sleeves, then place Teflon caps over the ends of the sleeves to be sent to the laboratory. If sample sleeves are not being used, spoon soil from the core barrel directly into a sample jar.

    To collect a composite sample, sample sleeves are not needed; rather, soil from each of the intervals to be composited should be transferred into a stainless steel bowl and homogenized prior to filling a sample jar.
7. After the sleeve or jar is capped, attach a sample label and custody seal to the jar and immediately place it into a Blue Ice®-packed cooler. Samples to be analyzed for radionuclides do not commonly require cooling.

8. See Chapter 5 for details on preparing samples and coolers for sample shipment.

9. Any soil left over from the sampling should be containerized in a DOT-approved 55-gal drum. Prior to leaving the site, all waste drums should be sealed, labeled, and handled appropriately (see Chapter 7).

10. Finally, the coordinates of the sampling point should be surveyed in by a professional surveyor to preserve the exact sampling location.

Open-Tube Sampler Method. For collecting shallow soil samples for lithology description, the open-tube sampler is recommended. This tool is comprised of an open-core barrel, extension rod, and T-bar Handle (Figure 3.19). Sampling tubes are available with a step to allow the sampler to use his/her weight to force the sampler into the ground. Since the sampling tube is open on one side, the soil lithology can be described without removing the sample from the tube. This tool works most effectively in moist nongravelly soils.

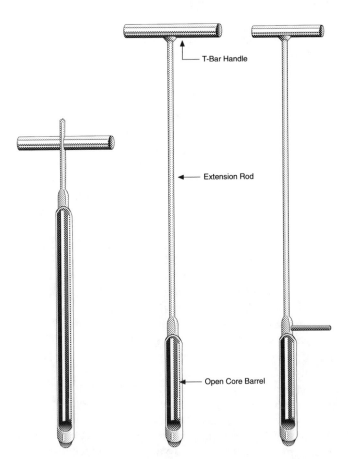

Figure 3.19. Three types of Open-Tube Samplers used to collect shallow soil samples.

For a large sampling program, three people are sufficient for this procedure. One is needed for sample collection and description; a second is needed for health and safety and quality control; and a third is needed for miscellaneous tasks such as waste management, and equipment decontamination.

The following equipment and procedure can be used to collect shallow soil samples for lithology description:

1. open-tube sampler and extension rods
2. stainless steel knife
3. sample labels
4. sample logbook
5. permanent ink marker
6. health and safety screening instruments
7. health and safety clothing
8. DOT-approved 55-gal waste drum
9. sample table
10. plastic sheeting
11. aluminum foil
12. plastic waste bags

Sampling Procedure

1. In preparation for sampling, read the introduction to Intrusive Sampling Methods (page 55) to confirm that all necessary preparatory work has been completed, including: obtaining property access agreements; meeting health and safety, decontamination, and waste disposal requirements; and calibrating all health and safety and sampling equipment.
2. Cut a 1-ft diameter hole in the center of the plastic sheeting, and center the hole over the sampling point. The purpose of this sheeting is to help prevent the spread of contamination.
3. If the tube sampler has a step, place one foot on the step and use your body weight to push the sample tube into the ground. Otherwise, use the T-bar handle to push the sampler into the ground.
4. Remove the sampler from the ground by pulling upward on the T-bar. To avoid injury, be certain to keep your back straight and lift with your legs.
5. Lay the sampler on the sampling table underlain by a piece of aluminum foil. Since the sampling tube is open-sided, the soil can be described without removing it from the tube. When describing the lithology, it is recommended that a knife be used to slice open the sample, to reveal the sample texture.
6. If the sample is to be archived, remove the soil core through the open side of the sample tube. Wrap the core in aluminum foil, then place it in a core box. Mark the box with the name of the sampler, sampling time, date, location, and depth, and seal it shut with custody tape.
7. To collect a deeper sample from the same hole, attach an extension rod to a clean sampling tube and repeat the above procedure.

8. Any soil left over from the sampling should be containerized in a DOT-approved 55-gal drum. Prior to leaving the site, all waste drums should be sealed, labeled, and handled appropriately (see Chapter 7).
9. Finally, the coordinates of the sampling point should be surveyed in by a professional surveyor to preserve the exact sampling location.

Split-Tube or Solid-Tube Method. The Split-Tube or Solid-Tube Method is very similar to the Slide Hammer Coring Method, with the exception being that a drill rig is used to beat the sampler into the ground. These samplers are composed of a split or solid sample tube, hardened shoe, soil catcher, and ball check (Figure 3.20). These samplers are available in two standard sizes, where the tubes are either 18- or 24-in. in length and have an outside diameter (O.D.) of 2- or 3-in.

If the sampler is being used to collect samples for laboratory analysis, it should be made of stainless steel. Using stainless steel sample liners to line the

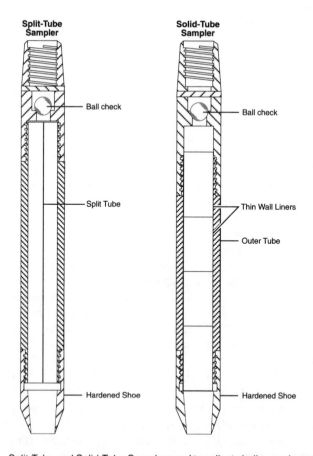

Figure 3.20.   Split-Tube and Solid-Tube Sampler used to collect shallow or deep soil samples.

sample tube is not a necessity; however, they are recommended when analyzing samples for volatile organics. Without sample liners, soil must be extracted from the tube and transferred into a sample jar using a stainless steel spoon. In this procedure, volatile organics can be lost into the ambient air. In contrast, sample liners can be quickly removed from the barrel, sealed with airtight Teflon caps, labeled, custody-sealed, then shipped to the laboratory for analysis. When samples are being collected for analyses other than volatile organics, or for lithology description, sample liners are not needed.

For most sampling programs, three people are sufficient for this sampling procedure in addition to the drill rig operators. One is needed for sample collection, lithology description, labeling, and documentation; a second is needed for health and safety and quality control; and a third is needed for miscellaneous tasks such as waste management, and equipment decontamination.

The following equipment and procedure can be used to collect shallow soil samples for chemical and/or radiological analysis:

1. stainless steel split-tube or solid-tube sampler
2. stainless steel sample sleeves
3. Teflon end-caps for stainless steel sleeves
4. auger drill rig with slide hammer
5. stainless steel bowl
6. stainless steel spoon
7. stainless steel knife
8. soil sample jars
9. sample labels
10. cooler packed with Blue Ice®
11. trip blank and coolant blank
12. sample logbook
13. chain-of-custody forms
14. chain-of-custody seals
15. permanent ink marker
16. health and safety screening instruments
17. health and safety clothing
18. DOT-approved 55-gal waste drum
19. sampling table
20. plastic sheeting
21. plastic waste bags

Sampling Procedure

1. In preparation for sampling, read the introduction to Intrusive Sampling Methods (page 55) to confirm that all necessary preparatory work has been completed, including: obtaining property access agreements; meeting health and safety, decontamination, and waste disposal requirements; and calibrating all health and safety and sampling equipment.

2. Cut a 1-ft diameter hole in the center of the plastic sheeting, and center the hole over the sampling point. The purpose of this sheeting is to help prevent the spread of contamination.
3. Have the drillers back the drill rig up to the sampling location, carefully raise the mast, then auger down to the top of the desired sampling interval.
4. Attach the sampler to a length of A-rod and lower it down the inside of the augers. Using the drill rig hammer, beat the sampler into the ground. Record the blow count in a sample logbook.
5. After removing the A-rod from the hole, detach the split-tube or solid-tube from the sampler.
6. To collect a grab sample using a split tube, break the tube open to reveal the sample sleeves. Using a stainless steel knife, separate the individual sleeves, then place Teflon caps over the ends of those to be sent to the laboratory for analysis. If sample sleeves are not being used, spoon soil from the split tube directly into a sample jar.

   To collect a grab sample using a solid tube, slide the sample sleeves out of one end of the solid tube. Using a stainless steel knife, separate the individual sleeves, then place Teflon caps over the ends of those to be sent to the laboratory for analysis. If sample sleeves are not being used, spoon soil from the solid tube directly into a sample jar.

   To collect a composite sample, there is no need to use sample sleeves. Rather, soil from each of the intervals to be composited should be transferred into a stainless steel bowl and homogenized prior to filling a sample jar.
7. After a sleeve or jar is capped, attach a sample label and custody seal, and immediately place it into a Blue Ice®-packed cooler. Samples to be analyzed for radionuclides do not commonly require cooling.
8. See Chapter 5 for details on preparing samples and coolers for sample shipment.
9. Any soil left over from the sampling should be containerized in a DOT-approved 55-gal drum. Prior to leaving the site, all waste drums should be sealed, labeled, and handled appropriately (see Chapter 7).
10. Finally, the coordinates of the sampling point should be surveyed in by a professional surveyor to preserve the exact sampling location.

Thin-Walled Tube Method. What is unique about this method is that the sample tube is hydraulically pushed into the ground using a drill rig, as opposed to being driven into the ground with a hammer. The advantage of pushing the sampler into the ground is that the soil is not artificially compacted in the sampling process. Consequently, the Thin-Walled Tube is the preferred method for collecting samples for geotechnical analysis. Some of the more common geotechnical tests run on soil samples include: porosity, hydraulic conductivity, specific gravity, grain size distribution, Atterberg Limits, compaction, consolidation, compression, and shear.

The Thin-Walled Tube Method utilizes a thin-walled (1/16-in.) sampling tube which has a standard O.D. of 3-in., and length that allows the collection of a 30-in. sample (Figure 3.21). There are four holes at the top of the tube which are used to connect the sampler to a sampling rod. Thin-walled tubes are available in either low carbon steel, or stainless steel. If only geotechnical analyses are to be performed on the sample, low carbon steel is acceptable. However, if the soil sample is to be tested for chemical or radiological composition, the sampler should be made of stainless steel.

For most sampling programs, three people are sufficient for this sampling procedure in addition to the drill rig operators. One is needed for sample collection, lithology description, labeling, and documentation; a second is needed for health and safety and quality control; and a third is needed for miscellaneous tasks such as waste management, and equipment decontamination.

The following equipment and procedure can be used to collect shallow soil samples for geotechnical testing:

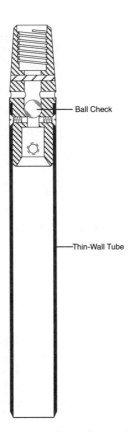

Ball Check

Thin-Wall Tube

Figure 3.21. Thin-Walled Tube Sampler used to collect shallow or deep soil samples.

1. thin-walled tube sampler
2. sample tube end-caps
3. auger drill rig
4. sampling knife
5. sample labels
6. sample logbook
7. chain-of-custody forms
8. chain-of-custody seals
9. permanent ink marker
10. health and safety screening instruments
11. health and safety clothing
12. DOT-approved 55-gal waste drum
13. sampling table
14. plastic sheeting
15. paraffin wax
16. plastic waste bags

Sampling Procedure

1. In preparation for sampling, read the introduction to Intrusive Sampling Methods (page 55) to confirm that all necessary preparatory work has been completed, including: obtaining property access agreements; meeting health and safety, decontamination, and waste disposal requirements; and calibrating all health and safety and sampling equipment.
2. Cut a 1-ft diameter hole in the center of the plastic sheeting, and center the hole over the sampling point. The purpose of this sheeting is to help prevent the spread of contamination.
3. Have the drillers back the drill rig up to the sampling location, carefully raise the mast, then auger down to the top of the desired sampling interval.
4. Attach the sampler to a length of A-rod and lower it down the inside of the augers. Using the drill rig, hydraulically push the sampler into the ground.
5. After removing the A-rod from the hole, detach the sample tube. Using a sampling knife, shave approximately 0.5-in. of soil from each end of the tube. Fill the space with melted paraffin wax, then place a cap over each end of the tube. The purpose of the wax is to prevent the shifting of soil in the tube during shipment to the geotechnical laboratory.
6. Attach a sample label to the tube, then place custody seals over each end cap.
7. When transporting samples to the geotechnical laboratory, sample tubes should be carried in a vertical position to preserve the soil compaction characteristics. If samples must be shipped, it is best to mark the sample box as "FRAGILE." Also, denote on the outside of the box which end is "UP." If no chemical analyses are being performed, there is no need to chill the sample.
8. Any waste left over from the sampling should be containerized in a DOT-

approved 55-gal drum. Prior to leaving the site, all waste drums should be sealed, labeled, and handled appropriately (see Chapter 7).

9. Finally, the coordinates of the sampling point should be surveyed in by a professional surveyor to preserve the exact sampling location.

Deep Soil Sampling

Soil samples collected at depths greater than 5 ft are generally referred to as "deep." These samples are most commonly collected by driving a split-tube or solid-tube sampler (Figure 3.20) or hydraulically pushing a thin-walled tube (Figure 3.21) into the ground with the assistance of an auger drill rig. These sampling methods are similar to those described for Shallow Soil Sampling (page 81) except samples are collected from deeper intervals.

Table 3.2 summarizes the effectiveness of the two recommended sampling methods. A number "1" in the table indicates that a particular procedure is most effective in collecting samples for a particular laboratory analysis, sample type, or sampling depth. A number "2" indicates that the procedure is acceptable, but less effective, while an empty cell indicates that the procedure is not recommended. For example, Table 3.2 indicates that the split-tube and solid-tube samplers are both effective in collecting soil samples for volatile organics and all other chemical analyses, while the thin-walled tube sampler is most effectively used to collect geotechnical samples. The following sections provide further details and SOPs for these deep soil sampling methods.

Split-Tube or Solid-Tube Method. The Split-Tube or Solid-Tube Method used to collect deep soil samples is identical to that described for shallow soil sampling (see page 81), with the exception being that samples are collected from a depth greater than 5 ft.

Table 3.2. Evaluation Table for Deep Soil Sampling Methods

	Laboratory Analyses								Sample Type			Sampling Depth	
	Volatiles	Semi-Volatiles	Primary Metals	Pesticides	PCBs	TPH	Radionuclides	Geotechnical	Grab	Composite (Vertical)	Composite (Areal)	Deep (>5 ft.)	Lithology Description
Split-Tube/Solid-Tube	1/1	1/1	1/1	1/1	1/1	1/1	1/1		1/1	1/2	2/2	1/1	1/2
Thin-Walled Tube								1	1			1	

1 = Preferred Method
2 = Acceptable Method
Empty Cell = Method Is Not Recommended

Thin-Walled Tube Method. The Thin-Walled Tube Method used to collect deep soil samples is identical to the procedure described for shallow soil sampling (see page 81), with the exception being that samples are collected from a depth greater than 5 ft.

Sediment Sampling

The following section provides the reader with guidance on selecting sediment sampling methods for remedial investigation studies. The criteria used in selecting the most appropriate method include the analyses to be performed on the sample, the type of sample being collected (grab or composite), and the sampling depth. SOPs have been provided for each of the methods to facilitate implementation. The following sediment characterization strategies are provided as a supplement to the DQO process outlined in Chapter 2.

One objective of an initial site characterization study should be to determine if contamination is present in the sediment of nearby surface water units such as streams, rivers, surface and storm sewer drainages, ponds, lakes, and/or retention basins. It is particularly important to characterize the sediment in streams, rivers, and other surface water drainages since they provide avenues for rapid contaminant migration, and provide points where receptors are readily exposed to contamination. Ponds, lakes, and retention basins do not provide the same opportunity for rapid contaminant migration; however, they similarly provide exposure points for receptors.

When performing an initial site characterization study (Stage 1), one grab sediment sample is commonly collected from near the center of each surface water and storm sewer drainage, sump, and retention pond at the site. These samples are collected from biased sampling locations which are downstream from potential sources of contamination and are commonly collected from the 0.0- to 0.5-ft BGS sampling interval. One upstream sediment sample should also be collected from surface water and storm sewer drainages to assist in defining background sediment chemistry (Figure 3.22).

The objective in collecting a sediment sample from each surface water and storm sewer drainage is to determine if contaminants are migrating offsite. If contaminants are identified at any of the Stage 1 sampling points, one should consider collecting additional samples as part of a Stage 2 sampling effort, including (Figure 3.22):

- deeper samples from the same locations to define the depth of contamination,
- upstream samples to assist in locating the contaminant source,
- downstream samples to define the extent of contamination, and
- cross-channel samples to define where in the channel the contaminant is concentrating (Figure 3.23).

The downstream sampling should include collecting samples from nearby streams, ponds, and/or lakes which receive sediment from the contaminated drainage. These samples are commonly collected from biased sampling loca-

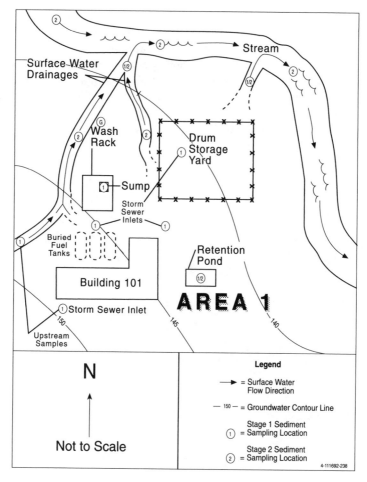

Figure 3.22. Example of a Stage 1 and Stage 2 sediment sampling effort.

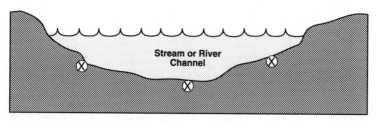

Ⓧ = Sediment Sampling Location

Figure 3.23. Cross-channel sediment sampling locations.

tions, such as just downstream from where each surface water drainage enters a stream (Figure 3.22), and near the inflow and outflow points in a pond or lake (Figure 3.24). If the contaminants of concern do not include volatile organics, initial characterization of ponds and lakes can be performed by collecting an areal composite sample across the lake (Figure 3.24). The number of samples required to complete the characterization is dependent on the DQOs, as discussed on page 33.

Most retention ponds are not watertight, and can slowly release contami-

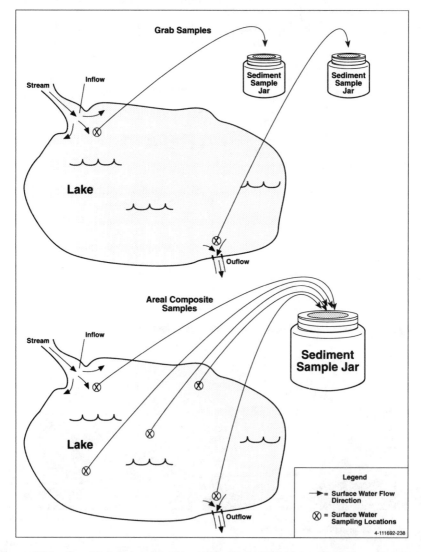

Figure 3.24.   Possible initial sediment sampling strategies for lakes or ponds.

nated water into the surrounding formation. For this reason, it is important to characterize the sediment held in these ponds. Since retention ponds are typically small in size, there is rarely a need to collect samples from more than one location.

The following sections present preferred sampling methods and procedures for collecting sediment samples from streams, rivers, and surface water drainages; and ponds, lakes, and retention basins. Sampling tools used to collect samples for laboratory analysis should be made of Teflon and/or stainless steel.

Stream, River, and Surface Water Drainage Sampling

Although a number of sophisticated sampling devices are available to collect sediment samples, not all of these tools are effective in collecting samples through the shallow, fast-moving water which typifies streams, rivers, and surface water drainages. The methods that have proved to be the most effective when sampling these environments include the Scoop or Dipper Method, Slide Hammer Method, and Box Sampler Method.

Of these three methods, the Scoop or Dipper Method is the easiest to implement since it simply involves pushing the sampler into the sediment, then transferring the sediment into a sample jar. This technique is only effective in collecting samples of the top 0.5 ft of sediment at locations where the water depth is less than 2 ft.

The Slide Hammer Method involves beating a sampling tube into the sediment. This technique removes a core of sediment for analytical testing. Since the sampling tube is typically lined with sampling sleeves, either individual sleeves can be sent to the laboratory as grab samples, or the soil can be removed from the sleeves and composited prior to filling a sample jar.

The Box Sampler Method utilizes a spring-loaded sample box attached to a sampling pole to collect grab sediment samples. This method involves pushing the sampling box into the sediment, and releasing the spring-loaded sample jaws. After the sampler is retrieved, sediment from the sampling box is transferred into a sample jar. Similar to the Scoop or Dipper Method, this method is only effective in collecting samples from the top 0.5 ft of sediment.

Depending on the depth of the water overlying the sediment, samplers may need a pair of waders, a raft, or a boat to access the sampling point. If waders are used, the sampler should face upstream while collecting the sample to assure that the sampler's boots do not contaminate the sample. Similarly, samples should be collected from the upstream side of the raft or boat. As a general rule, the sample located farthest downstream should always be the first to be collected. Sampling should then proceed upstream. By collecting samples in this manner, any sediment disturbed by the samplers will not contaminate downstream sampling points.

Table 3.3 summarizes the effectiveness of each of the three recommended sampling methods. A number "1" in the table indicates that a particular proce-

Table 3.3. Evaluation Table for Stream, River, and Surface Water Drainage Sediment Sampling Methods

	Laboratory Analyses								Sample Type			Sampling Depth		Lithology Description
	Volatiles	Semi-Volatiles	Primary Metals	Pesticides	PCBs	TPH	Radionuclides	Geotechnical	Grab	Composite (Vertical)	Composite (Areal)	Surface (0-0.5 ft.)	Shallow (0-3.0 ft.)	
Scoop or Dipper	2	2	2	2	2	2	2		2		2	2		2
Slide-Hammer	1	1	1	1	1	1	1		1	1	1	2	1	2
Box Sampler	2	1	1	1	1	1	1		1		1	1		1

1 = Preferred Method
2 = Acceptable Method
Empty Cell = Method Is Not Recommended

dure is most effective in collecting samples for a particular laboratory analysis, sample type, or sampling depth. A number "2" indicates that the procedure is acceptable, but less effective, while an empty cell indicates that the procedure is not recommended. For example, Table 3.3 indicates that the Slide Hammer Method is most effective in collecting sediment samples for volatile organics. While the Scoop or Dipper, and Box Sampler are acceptable sampling methods for volatile organic analysis, they are less effective than the Slide Hammer Method.

Scoop or Dipper Method. The Scoop or Dipper is the simplest sampling tool for collecting grab sediment samples from streams, rivers, and surface water drainages, and is available in many shapes and sizes (Figure 3.25). This method involves lowering a Teflon or stainless steel scoop or dipper by hand through the surface water and pushing the sampler deep into the underlying sediment. As the sampler is retrieved, the sediment is transferred into a sample jar. This technique is generally effective in collecting samples of the top 0.5 ft of sediment, at locations where the water is less than 2 ft in depth.

This method is most effective when collecting samples from water bodies with relatively slow flow velocities, since a significant amount of the finer grained sediment tends to be lost when sampling higher energy environments. The dipper typically works more effectively than the scoop at preventing the loss of fine grained sediment when retrieving the sample; however, since the dipper has no cutting edge it is only effective in sampling soft sediment. The Scoop or Dipper Method is most commonly used for preliminary remedial investigation studies. If contaminants are identified during the preliminary sampling, the Slide Hammer (page 106) or Box Sampler (page 109) Methods should be considered to more accurately define the distribution of contaminants.

Sampling Scoops

Sampling Dipper

Figure 3.25. Various types of scoops and a dipper for sediment sampling.

If a grab sample is collected, sediment is transferred from the sampler directly into a sample jar. If a composite sample is collected, the sediment to be composited is transferred into a stainless steel bowl and homogenized with a stainless steel spoon prior to filling a sample jar.

For most sampling programs, four people are sufficient for this sampling procedure. Two are needed for sample collection, labeling, and documentation; a third is needed for health and safety, and quality control; and a fourth is needed for waste management, and equipment decontamination.

The following equipment and procedure can be used to collect sediment samples for chemical and/or radiological analysis:

1. Teflon or stainless steel scoop or dipper
2. stainless steel bowl
3. stainless steel spoon
4. sample jars
5. sample labels
6. cooler packed with Blue Ice®
7. trip blank and coolant blank
8. sample logbook
9. chain-of-custody forms
10. chain-of-custody seals
11. permanent ink marker
12. health and safety screening instruments

13. health and safety clothing
14. DOT-approved 55-gal waste drum
15. sampling table
16. plastic waste bags

Sampling Procedure

1. In preparation for sampling, read the introduction to Intrusive Sampling Methods (page 55) to confirm that all necessary preparatory work has been completed, including: obtaining property access agreements; meeting health and safety, decontamination, and waste disposal requirements; and calibrating all health and safety and sampling equipment.
2. Approach the sampling point from downstream, being careful not to disturb the underlying sediment.
3. Push the scoop or dipper firmly downward into the sediment, then lift upward. Quickly raise the sampler out of the water in an effort to reduce the amount of sediment lost to the water current. If a grab sample is being collected, transfer the sediment from the scoop or dipper directly into a sample jar. If a composite sample is being collected, transfer the sediment from each composite interval or location into a stainless steel bowl and homogenize with a stainless steel spoon prior to filling a sample jar.
4. After the jar is capped, attach a sample label and custody seal to the jar and immediately place it into a Blue Ice®-packed cooler. Samples to be analyzed for radionuclides do not commonly require cooling.
5. See Chapter 5 for details on preparing sample jars and coolers for sample shipment.
6. Any waste material left over from the sampling should be containerized in a DOT-approved 55-gal drum. Prior to leaving the site, all waste drums should be sealed, labeled, and handled appropriately (see Chapter 7).
7. Finally, the coordinates of the sampling point should be surveyed in by a professional surveyor to preserve the exact sampling location.

Slide-Hammer Method. The Slide Hammer is an effective tool for collecting core samples of stream and river sediment. This tool is comprised of a stainless steel core barrel, extension rod, and slide hammer (Figure 3.26). Stock core barrels have an I.D. of 2- or 2.5-in., and are 2- to 3-ft in length. However, the core barrel can be special-ordered to meet project specific volume requirements. The top of the barrel is threaded so that it can be screwed into an extension rod to allow sampling through deeper water.

The barrel is constructed to accept sample liners which are commonly used to facilitate the removal of sediment from the barrel without disturbing the sample. The use of liners is not a necessity; however, they are recommended when grab samples are to be analyzed for volatile organics. Without sample liners, sediment must be extracted from the barrel and transferred into a sample jar. In this process, volatile organics can be lost into the ambient air. In

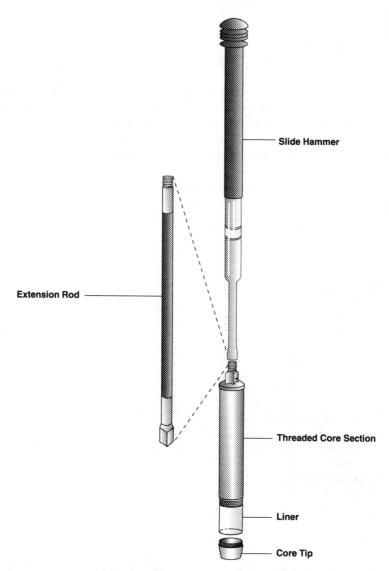

Figure 3.26.   Slide Hammer used to collect sediment core samples.

contrast, sample liners can be quickly removed from the barrel and sealed with airtight Teflon caps. After labeling and custody sealing the liners, they can be shipped directly to the laboratory for analysis. When collecting a composite sample, sediment from the intervals or locations to be composited is transferred into a stainless steel bowl and homogenized prior to filling a sample jar.

Extension rods can be ordered in various lengths to allow sampling through various depths of water. The rods are screwed into the core barrel at one end, and into the slide hammer at the other end. The slide hammer is used to beat the sampler into the sediment. The hammer is available in various weights to accommodate the needs of the sampler. If samples are to be collected from locations where water depths exceed several feet, a raft or boat will be required to assist the sampling procedure.

For most sampling programs, four people are sufficient for this sampling procedure. Two are needed for sample collection, labeling, and documentation; a third is needed for health and safety and quality control; and a fourth is needed for waste management and equipment decontamination. If a raft or boat is used to assist the sampling procedure, at least one additional person will be needed.

The following equipment and procedure can be used to collect sediment samples for chemical and/or radiological analysis:

1. slide hammer and extension rods
2. stainless steel sample sleeves
3. Teflon end-caps for sample sleeves
4. stainless steel bowl
5. stainless steel spoon
6. stainless steel knife
7. sample jars
8. sample labels
9. cooler packed with Blue Ice®
10. trip blank and coolant blank
11. sample logbook
12. chain-of-custody forms
13. chain-of-custody seals
14. permanent ink marker
15. health and safety screening instruments
16. health and safety clothing
17. waders, raft, or boat
18. sampling table
19. DOT-approved 55-gal waste drum
20. plastic waste bags

Sampling Procedure

1. In preparation for sampling, read the introduction to Intrusive Sampling Methods (page 55) to confirm that all necessary preparatory work has been completed, including: obtaining property access agreements; meeting health and safety, decontamination, and waste disposal requirements; and calibrating all health and safety and sampling equipment.
2. Approach the sampling point from downstream, being careful not to disturb the underlying sediment.
3. Lower the sampler through the water, then beat the core barrel to the desired depth, and record the blow counts in a sample logbook.
4. Remove the core barrel from the hole by either rocking it from side to side several times before lifting, or reverse beating the sampler from the hole.

5. To collect a grab sample, unscrew the core barrel from the sampler and slide the sample sleeves out onto the sampling table. Using a stainless steel knife, separate the sample sleeves, then place Teflon caps over the ends of the sleeves to be sent to the laboratory. If sampling sleeves are not being used, spoon sediment from the core barrel directly into a sample jar.

   To collect a composite sample, sample sleeves are not needed; rather, sediment from each of the intervals to be composited should be transferred into a stainless steel bowl and homogenized prior to filling a sample jar.

6. Attach a sample label and custody seal to the sample sleeve or jar, and place it in a Blue Ice®-packed cooler. Samples to be analyzed for radionuclides do not commonly require cooling.

7. See Chapter 5 for details on preparing sample jars and coolers for sample shipment.

8. Any sediment left over from the sampling should be containerized in a DOT-approved 55-gal drum. Prior to leaving the site, the waste drum should be sealed, labeled, and handled appropriately (see Chapter 7).

9. Finally, the coordinates of the sampling point should be surveyed in by a professional surveyor to preserve the exact sampling location.

Box Sampler Method. The stainless steel box sampler is a very effective tool for collecting grab samples of stream and river sediment. This sampler is composed of a sample box, spring-loaded sample jaws, and a pole containing a spring release mechanism (Figure 3.27). After the sample box is pushed deeply into the sediment, the jaws are released to seal off the bottom of the box. The closed box is then retrieved from the water. The advantage of this sampling method is that fine grained sediment is not stripped from the sample as it is removed from the water.

For most sampling programs, four people are sufficient for this sampling procedure. Two are needed for sample collection, labeling, and documentation; a third is needed for health and safety and quality control; and a fourth is needed for waste management and equipment decontamination. If a raft or boat is used to assist the sampling procedure, at least one additional person will be needed.

The following equipment and procedure can be used to collect sediment samples for chemical and/or radiological analysis:

1. stainless steel box sampler
2. stainless steel bowl
3. stainless steel spoon
4. sample jars
5. sample labels
6. cooler packed with Blue Ice®
7. trip blank and coolant blank
8. sample logbook
9. chain-of-custody forms
10. chain-of-custody seals

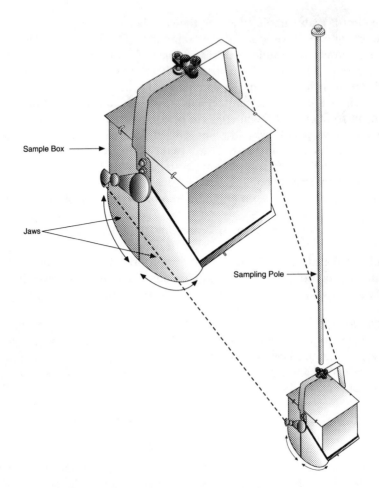

Figure 3.27. Box Sampler used to collect sediment samples.

11. permanent ink marker
12. health and safety screening instruments
13. health and safety clothing
14. waders, raft, or boat
15. sampling table
16. DOT-approved 55-gal waste drum
17. plastic waste bags

Sampling Procedure

1. In preparation for sampling, read the introduction to Intrusive Sampling Methods (page 55) to confirm that all necessary preparatory work has been completed, including: obtaining property access agreements; meeting

health and safety, decontamination, and waste disposal requirements; and calibrating all health and safety and sampling equipment.

2. Approach the sampling point from downstream, being careful not to disturb the underlying sediment.

3. Hold the sampling pole so the open sampler jaws are positioned several inches above the surface of the sediment, then firmly thrust the sampler downward. Depress the button at the top of the sampling pole to release the spring-loaded jaws.

4. If a grab sample is being collected, transfer the sediment from the box sampler directly into a sample jar. If a composite sample is being collected, transfer the sediment from the locations to be composited into a stainless steel bowl and homogenize with a stainless steel spoon prior to filling a sample jar.

5. After the jar is capped, attach a sample label and custody seal to the jar and immediately place it into a Blue Ice®-packed cooler. Samples to be analyzed for radionuclides only do not commonly require cooling.

6. See Chapter 5 for details on preparing sample jars and coolers for sample shipment.

7. Any sediment left over from the sampling should be containerized in a DOT-approved 55-gal drum. Prior to leaving the site, all waste drums should be sealed, labeled, and handled appropriately (see Chapter 7).

8. Finally, the coordinates of the sampling point should be surveyed in by a professional surveyor to preserve the exact sampling location.

Pond, Lake, and Retention Basin Sampling

The sampling methods which have proved to be the most effective when sampling sediment from ponds, lakes, and retention basins include the Scoop or Dipper, Slide Hammer, Box Sampler, and Dredge Sampler. The Scoop or Dipper, Slide Hammer, and Box Sampler Methods are all effective when collecting sediment samples from around the edges of ponds, lakes, and retention basins, where the water is relatively shallow. On the other hand, the Box Sampler and Dredge Sampler, used in combination with a wire line, are effective methods for collecting sediment samples through deep water.

Of these four methods, the Scoop or Dipper Method is the easiest to implement since it simply involves lowering a Teflon or stainless steel scoop or dipper by hand through the surface water and pushing the sampler deep into the underlying sediment. As the sampler is retrieved, the sediment is transferred into a sample jar. This technique is generally effective in collecting samples of the top 0.5 ft of sediment at locations where the water depth is less than 2 ft.

The Slide Hammer Method involves beating a sampling tube into the sediment. This technique removes a core of sediment for analytical testing. Since the sampling tube is typically lined with sampling sleeves, either individual sleeves can be sent to the laboratory as grab samples, or the soil can be removed from the sleeves and composited prior to filling a sample jar. Of the four sampling methods, only the Slide Hammer is effective in collecting sediment samples deeper than the top 0.5 ft. When used in combination with extension rods, the Slide Hammer can be used to collect a 2- to 3-ft sediment core through 10- to 15-ft of water.

The Box Sampler Method utilizes a spring-loaded sample box attached to either a sampling pole or wire line to collect grab sediment samples. This method involves pushing the Box Sampler into the sediment, and releasing the spring-loaded sample jaws. After the sampler is retrieved, sediment from the sampling box is transferred into a sample jar. This method is only effective in collecting samples from the top 0.5 ft of sediment.

The Dredge Sampler Method is composed of two jaws connected by a lever, and is lowered through the water using a wire line. As the sampler is quickly lowered through the water, the jaws open and embed themselves in the underlying sediment. As the wire line is raised, the jaws close to capture a sample. Sediment from the dredge is then transferred into a sample jar using a stainless steel spoon. Similar to the Scoop or Dipper and Box Sampler, this method is only effective in collecting samples from the top 0.5 ft of sediment.

Depending on the depth of the water overlying the sediment, samplers will need either a pair of waders, a raft, or a boat to access the sampling point. Table 3.4 summarizes the effectiveness of each of the four recommended sampling methods. A number "1" in the table indicates that a particular procedure is most effective in collecting samples for a particular laboratory analysis, sample type, or sampling depth. A number "2" indicates that the procedure is acceptable but less effective, while an empty cell indicates that the procedure is not recommended. For example, Table 3.4 indicates that the Slide Hammer is the most effective method for collecting sediment samples for volatile organic

Table 3.4. Evaluation Table for Pond, Lake, and Retention Pond Sediment Sampling Methods

| | Laboratory Analyses | | | | | | | | Sample Type | | | Sampling Depth | | |
	Volatiles	Semi-Volatiles	Primary Metals	Pesticides	PCBs	TPH	Radionuclides	Geotechnical	Grab	Composite (Vertical)	Composite (Areal)	Surface (0-0.5 ft.)	Shallow (0.0-3.0 ft.)	Lithology Description
Scoop or Dipper	2	2	2	2	2	2	2		2		2	2		2
Slide Hammer	1	1	1	1	1	1	1		1	1	1		1	1
Box Sampler	2	1	1	1	1	1	1		1		1	1		1
Dredge Sampler	2	1	1	1	1	1	1		1		1	1		1

1 = Preferred Method
2 = Acceptable Method
Empty Cell = Method Is Not Recommended

analysis. While the Scoop or Dipper, Box Sampler, and Dredge Sampler are acceptable sampling methods for volatile organic analysis, they are less effective than the Slide Hammer Method.

Scoop or Dipper Method. The Scoop or Dipper Method used to collect sediment samples from ponds, lakes, and retention basins is identical to the procedure described for streams, rivers, and surface water drainage sampling (see page 104).

Slide Hammer Method. The Slide Hammer Method used to collect sediment samples from ponds, lakes, and retention basins is identical to the procedure described for streams, rivers, and surface water drainage sampling (see page 106).

Box Sampler Method. The Box Sampler Method used to collect sediment samples from ponds, lakes, and retention basins is identical to the procedure described for streams, rivers, and surface water drainage sampling (see page 109), with the following modifications:

- A wire line may be used to lower the sampler through the water, as opposed to the sampling pole.
- Add "wire line" to the equipment list.
- Modify Step 3 to read "Hold the sampling pole so the open sampler jaws are positioned several inches above the surface of the sediment, then firmly thrust the sampler downward. Depress the button at the top of the sampling pole to release the spring-loaded jaws. When using a cable to lower the sampler, as opposed to a sampling pole, allow the sampler to free fall through the last 5- to 10-ft of water to assure that the sampler jaws get deeply embedded into the sediment. Then slide a trip weight down the cable to trip the spring-loaded jaws.

Dredge Sampler Method. The Bottom Dredge sampler is a very common and effective tool in collecting grab samples of sediment in ponds, lakes, and retention basins. This sampler is composed of two jaws connected by a lever (Figure 3.28). To use this sampling method, the dredge is attached to a cable and allowed to free-fall through the water to assure that the sampler jaws deeply embed themselves into the underlying sediment. As the cable is retrieved, the jaws to the sampler are forced closed. As with the Box Sampler, this technique is effective in preventing the loss of fine grained sediment as the sample is retrieved.

For most sampling programs, five people are sufficient for this sampling procedure. Two are needed for sample collection, labeling and documentation; a third is needed for health and safety and quality control; a fourth is needed for waste management and equipment decontamination; and a fifth is needed to maneuver the raft or boat.

The following equipment and procedure can be used to collect sediment samples for chemical and/or radiological analysis:

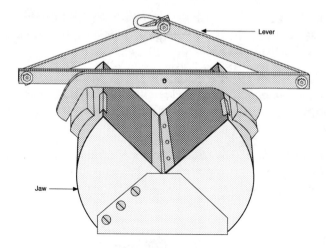

Figure 3.28. Dredge Sampler used to collect sediment samples.

1. stainless steel dredge sampler
2. wire line
3. stainless steel spoon
4. sample jars
5. sample label
6. cooler packed with Blue Ice®
7. trip blank and coolant blank
8. sample logbook
9. chain-of-custody forms
10. chain-of-custody seals
11. permanent ink marker
12. health and safety screening instruments
13. health and safety clothing
14. boat large enough for five people and sampling equipment
15. sampling table
16. DOT-approved 55-gal waste drum
17. plastic waste bags

Sampling Procedure

1. In preparation for sampling, read the introduction to Intrusive Sampling Methods (page 55) to confirm that all necessary preparatory work has been completed, including: obtaining property access agreements; meeting health and safety, decontamination, and waste disposal requirements; and calibrating all health and safety and sampling equipment.
2. Maneuver the raft or boat over the sampling location.
3. Lower the dredge sampler to the bottom of the pond or lake using a wire line. The faster the sampler is dropped, the deeper the sampler will be

embedded into the sediment. When the sampler hits bottom, allow the line to go slack for a few seconds, then retrieve.

4. If a grab sample is being collected, transfer the sediment from the dredge directly into a sample jar. If a composite sample is being collected, transfer the sediment to be composited into a stainless steel bowl and homogenize with a stainless steel spoon prior to filling a sample jar.

5. After the jar is capped, attach a sample label and custody seal to the jar and immediately place it into a Blue Ice®-packed cooler. Samples to be analyzed for radionuclides do not commonly require cooling.

6. See Chapter 5 for details on preparing sample jars and cooler for sample shipment.

7. Any sediment left over from the sampling should be containerized in a DOT-approved 55-gal drum. Prior to leaving the site, the waste drum should be sealed, labeled, and handled appropriately (see Chapter 7).

8. Finally, the coordinates of the sampling point should be surveyed in by a professional surveyor to preserve the exact sampling location.

Surface Water Sampling

The following section provides the reader with guidance on selecting surface water sampling methods for site characterization. The criteria used in selecting the most appropriate method include the analyses to be performed on the sample, the type of sample being collected (grab, composite, or integrated), and the sampling depth. SOPs have been provided for each of the methods to facilitate implementation. The following surface water characterization strategies are provided as a supplement to the DQO process outlined in Chapter 2.

One objective of an initial site characterization study should be to determine if contamination is present in nearby surface water units such as streams, rivers, surface and storm sewer drainages, ponds, lakes, and/or retention basins. It is particularly important to characterize the surface water in streams, rivers, and other surface water drainages since they provide avenues for rapid contaminant migration, and provide points where receptors are readily exposed to contamination. Ponds, lakes, and retention basins do not provide the same opportunity for rapid contaminant migration; however, they similarly provide exposure points for receptors.

When performing an initial site characterization study (Stage 1), one grab sample is commonly collected from each surface water and storm sewer drainage, sump, and retention pond at the site. These samples are collected from biased sampling locations which are downstream from potential sources of contamination. One upstream water sample should also be collected from surface water and storm sewer drainages to assist in defining background surface water chemistry (Figure 3.29).

The objective in collecting a sample from each surface water and storm sewer drainage is to determine if contaminants are currently migrating offsite. If contaminants are identified from the Stage 1 sampling, one should consider

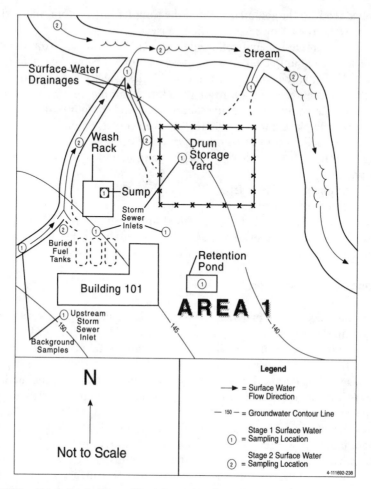

Figure 3.29. Example of a Stage 1 and Stage 2 surface water sampling effort.

collecting additional upstream samples to assist in locating the contaminant source, and downstream samples to further define the extent of contamination (Stage 2). The downstream sampling should include collecting samples from nearby streams, ponds, and/or lakes which receive water from the contaminated drainage. These samples are commonly collected from biased sampling locations, such as just downstream from where each surface water drainage enters a stream (Figure 3.29), and near the inflow and outflow points in a pond or lake (Figure 3.30). If the contaminants of concern do not include volatile organics, initial characterization of ponds and lakes can be performed by collecting an areal composite sample (Figure 3.30). The number of samples required to complete the characterization is dependent on the DQOs, as discussed on page 33.

Most retention ponds are not watertight, and can slowly release contami-

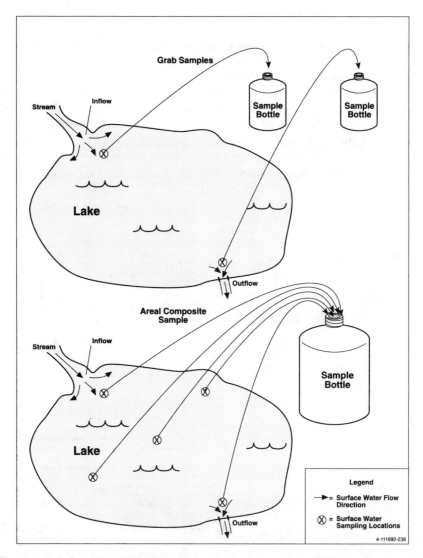

Figure 3.30.   Possible initial surface water sampling strategies for lakes or ponds.

nated water into the surrounding formation. For this reason, it is important to characterize the water held in these ponds. Since retention ponds are typically small in size, there is rarely a need to collect more than one surface water sample from each.

The depth at which surface water samples are collected should be based on the suspected concentration and density of the contaminants of concern. If Dense Non-Aqueous Phase Liquids (DNAPLs) are suspected, samples should be collected from the bottom of the water column. Otherwise, samples are most commonly collected from the top of the water column. If historical information leads the investigator to believe that contamination could be layered, due to varying specific gravities of the contaminants, one should consider collecting either a vertical composite sample, or several grab samples from different depth intervals (Figure 3.31).

When preparing to collect a surface water sample, a pre-sample should be collected in a glass jar, and should be analyzed for pH, temperature, and conductivity. Additional measurements which should be considered include turbidity, alkalinity, dissolved oxygen, and oxidation-reduction potential.

The following sections present preferred sampling methods and procedures for collecting surface water samples from streams, rivers, and surface water

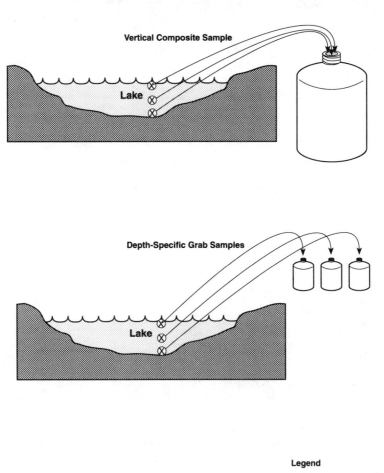

Figure 3.31.   Vertical grab or composite surface water sampling in a lake.

drainages, and ponds, lakes, and retention basins. Sampling tools used to collect samples for laboratory analysis should be made of Teflon, stainless steel, or glass.

Stream, River, and Surface Water Drainage Sampling

Although a number of sophisticated sampling devices are available to collect surface water samples, not all of these tools are effective in collecting samples in the shallow, fast-moving water which typify streams, rivers, and surface water drainages. Since the movement of water in these environments tends to naturally homogenize the water column, in most cases samples are collected from the water surface. However, when DNAPLs are suspected, deeper sampling can be performed.

The methods that have proved to be the most effective when sampling these environments include the Bottle Submersion Method, Dipper Method, Extendable Bottle Sampler Method, and Extendable Tube Sampler Method.

The Bottle Submersion, Dipper, Extendable Bottle Sampler, and Extendable Tube Sampler are all effective methods for collecting grab samples. Of these methods, the Bottle Submersion Method is the easiest to implement, since it simply involves submersing a sample bottle beneath the water surface. With this method, a telescoping extension rod is often used to hold the sample bottle as it is lowered into the water. The Dipper Method is also very implementable since it simply involves lowering a stainless steel dipper below the water surface, then transferring the water into sample bottles.

The Extendable Bottle Sampler Method utilizes a glass sample bottle attached to a vertical sampling pole. After lowering the bottle to the desired sampling depth, the seal ring is lifted, which allows the bottle to fill with water. When the bottle is full, the seal ring is depressed. The sampler is then retrieved and water is transferred into a sample bottle. This method is effective in collecting samples as deep as 5 ft below the water surface.

The Extendable Tube Sampler utilizes a sampling tube and a vertical extension rod. Similar to the Extendable Bottle Sampler, this method utilizes a check valve to open and close the sampling tube when the desired sampling depth is reached. After the sampling tube is retrieved, water is transferred into a sample bottle. This method is effective in collecting samples as deep as 5 ft below the water surface.

Any of the four methods can be used to collect areal, composite or integrated samples, where an areal composite sample is collected by compositing a number of grab samples from different locations, and an integrated sample involves collecting a portion of a sample from the same location several times over a period of time. Depth composite samples can be collected using either the Extendable Bottle Sampler, or Extendable Tube Sampler. A depth composite sample is collected by compositing several grab samples, each representing a different depth in the water column.

Table 3.5 summarizes the effectiveness of each of the four recommended sampling methods. A number "1" in the table indicates that a particular procedure is most effective in collecting samples for a particular laboratory analysis, sample type, or sampling depth. A number "2" indicates that the procedure is acceptable but less effective, while an empty cell indicates that the procedure is not recommended. For example, Table 3.5 indicates that Bottle Submersion is the preferred procedure for collecting samples for volatile organic analysis. While the Dipper, Extendable Bottle Sampler, and Extendable Tube Sampler are acceptable methods for volatile organic analysis, they are less desirable since the water from the sampler must be poured into sample bottles which tends to aerate the sample.

Whichever sampling method is selected, sample bottles for volatile organic analysis should always be the first to be filled. Bottles for the remaining parameters should be filled consistent with their relative importance to the sampling program. As a general rule, the sample located farthest downstream should always be the first to be collected. Sampling should then proceed upstream. By collecting samples in this manner, any sediment disturbed by the samplers will not contaminate downstream sampling points. Depending on the depth of the water, samplers may need a pair of waders, a raft, or a boat to access the sampling point. If waders are used, the sampler should face upstream while collecting the sample to assure that the waders do not contaminate the sample. Similarly, samples should be collected from the upstream side of a raft or boat.

Bottle Submersion Method. The Bottle Submersion Method is the simplest and one of the most commonly used surface water sampling method for collecting samples from streams, rivers, and surface water drainages. This method utilizes a water sample bottle, and an optional telescoping extension

Table 3.5. Evaluation Table for Stream, River, and Drainage Surface Water Sampling Methods

	Laboratory Analyses							Sample Type				Sampling Depth		
	Volatiles	Semi-Volatiles	Primary Metals	Pesticides	PCBs	TPH	Radionuclides	Grab	Composite (Verticle)	Composite (Areal)	Integrated	Surface (0.0-0.5 ft.)	Shallow (0.5-5.0 ft.)	Deep (>5.0 ft.)
Bottle Submersion	1	1	1	1	1	1	1	1		1	1	1		
Dipper	2	1	1	1	1	1	1	1		1	1	1		
Extendable Bottle Sampler	2	1	1	1	1	1	1	1	1	1	1	2	1	
Extendable Tube Sampler	2	1	1	1	1	1	1	1	1	1	1	2	1	

1 = Preferred Method
2 = Acceptable Method
Empty Cell = Method Is Not Recommended

rod with an adjustable beaker clamp (Figure 3.32). This method is only effective in collecting grab samples from the top few inches of the water column.

The following procedure can be used to collect grab surface water samples, and is written to include the use of a telescoping extension rod. If an extension rod is not available, the bottles can be lowered into the water by hand. Since this procedure is both simple and effective, it is recommended that initial surface water sampling be performed using either this method or the Dipper Method. If contaminants are identified from this preliminary sampling, the Extendable Bottle Sampler or Extendable Tube Sampler Methods should be considered to more accurately define the distribution of contaminants. For a summary of the effectiveness and limitations of this sampling method, see Table 3.5.

For most sampling programs, four people are sufficient for this sampling procedure. Two are needed for field testing, sample collection, labeling, and documentation; a third is needed for health and safety, and quality control; and a fourth is needed for miscellaneous tasks such as waste management, and equipment decontamination.

The following equipment and procedure can be used to collect surface water samples for chemical and/or radiological analysis:

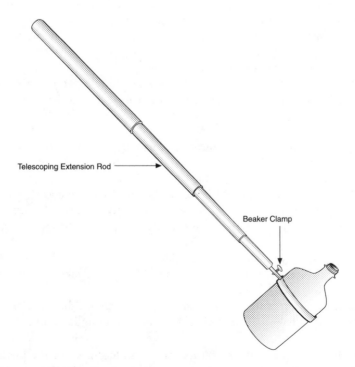

Telescoping Extension Rod

Beaker Clamp

Figure 3.32.   Telescoping extension rod used to assist the Bottle Submersion Method.

1. telescoping extension rod with adjustable beaker clamp
2. sample bottles
3. sample preservatives
4. pH, temperature, and conductivity meters
5. sample labels
6. cooler packed with Blue Ice®
7. trip blank and coolant blank
8. sample logbook
9. chain-of-custody forms
10. chain-of-custody seals
11. permanent ink marker
12. health and safety screening instruments
13. health and safety clothing
14. DOT-approved 55-gal waste drum
15. sampling table
16. plastic waste bags

Sampling Procedure

1. In preparation for sampling, read the introduction to Intrusive Sampling Methods (page 55) to confirm that all necessary preparatory work has been completed, including: obtaining property access agreements; meeting health and safety, decontamination, and waste disposal requirements; and calibrating all health and safety and sampling equipment.
2. Prior to sample collection, fill a clean glass jar with sample water, and measure the pH, temperature, and conductivity of the water. Record this information in a sample logbook.
3. Secure a clean sample bottle to the end of the telescoping extension rod using a beaker clamp. Remove the bottle cap just prior to sampling.
4. While standing on the bank of the stream or river, extend the rod out over the water and lower the sample bottle just below the water surface. If the analyses to be performed require the sample to be preserved, this should be performed prior to filling the sample bottle.
5. After the bottle is capped, attach a sample label and custody seal to the bottle and immediately place it into a Blue Ice®-packed cooler. Samples to be analyzed for radionuclides do not commonly require cooling.
6. See Chapter 5 for details on preparing sample bottles and coolers for sample shipment.
7. Containerize any waste water in a DOT-approved 55-gal drum. Prior to leaving the site, all waste drums should be sealed, labeled, and handled appropriately (see Chapter 7).
8. Finally, the coordinates of the sampling point should be surveyed in by a professional surveyor to preserve the exact sampling location.

Dipper Method. The Dipper Method is a simple but effective method for collecting grab samples of surface water from streams and rivers for both chemical and radiological analysis. With this method, a Teflon or stainless steel dipper is used to collect a water sample, then transfer it into a sample bottle (Figure 3.33). This method is only effective in collecting grab samples from the top few inches of the water column.

Since this procedure is both simple and effective, it is recommended that initial surface water sampling be performed using either this method or the Bottle Submersion Method. If contaminants are identified from this preliminary sampling, the Extendable Bottle Sampler or Extendable Tube Sampler Methods should be considered to more accurately define the distribution of contaminants. For a summary of the effectiveness and limitations of this sampling method, see Table 3.5.

For most sampling programs, four people are sufficient for this sampling procedure. Two are needed for field testing, sample collection, labeling, and documentation; a third is needed for health and safety and quality control; and a fourth is needed for waste management and equipment decontamination.

The following equipment and procedure can be used to collect surface water samples for chemical and/or radiological analysis:

1. Teflon or stainless steel dipper
2. sample bottles
3. sample preservatives
4. pH, temperature, and conductivity meters
5. sample labels
6. cooler packed with Blue Ice®
7. trip blank and coolant blank
8. sample logbook
9. chain-of-custody forms
10. chain-of-custody seals

Figure 3.33. Surface water sampling dipper.

11. permanent ink marker
12. health and safety screening instruments
13. health and safety clothing
14. DOT-approved 55-gal waste drum
15. sampling table
16. plastic waste bags

Sampling Procedure

1. In preparation for sampling, read the introduction to Intrusive Sampling Methods (page 55) to confirm that all necessary preparatory work has been completed, including: obtaining property access agreements; meeting health and safety, decontamination, and waste disposal requirements; and calibrating all health and safety and sampling equipment.
2. Prior to sample collection, fill a clean glass jar with sample water, and measure the pH, temperature, and conductivity of the water. Record this information in a sample logbook.
3. Extend the dipper out over the water and lower it just below the water surface, being careful not to disturb the underlying sediment.
4. When the dipper is full, carefully transfer the water into a sample bottle. If the analyses to be performed require the sample to be preserved, this should be performed prior to filling the sample bottle.
5. After the bottle is capped, attach a sample label and custody seal to the bottle and immediately place it into a Blue Ice®-packed cooler. Samples to be analyzed for radionuclides do not commonly require cooling.
6. See Chapter 5 for details on preparing sample bottles and coolers for sample shipment.
7. Containerize any waste water in a DOT-approved 55-gal drum. Prior to leaving the site, all waste drums should be sealed, labeled, and handled appropriately (see Chapter 7).
8. Finally, the coordinates of the sampling point should be surveyed in by a professional surveyor to preserve the exact sampling location.

Extendable Bottle Sampler Method. The Extendable Bottle Sampler is an effective tool for collecting grab samples as deep as 5 ft below the water surface. This sampler is composed of a glass sample bottle, vertical sampling pole, handle, and bottle seal ring (Figure 3.34).

This method and the Extendable Tube Sampler Method (page 127) are particularly effective for characterizing streams and rivers which are suspected of having stratified zones of contamination, since they can collect grab samples from various depth intervals within a water column. While this method is most effective in collecting grab samples, it can also be used to collect composite samples by combining a number of grab samples together in one sample bottle (see Figures 3.30 and 3.31). A composite sample should not be collected when the analyses to be performed on the sample include volatile organics, since

Figure 3.34. Extendable Bottle Sampler used to collect surface water samples.

volatilization is facilitated by the compositing process. For a summary of the effectiveness and limitations of this sampling method, see Table 3.5.

For most sampling programs, four people are sufficient for this sampling procedure. Two are needed for field testing, sample collection, labeling, and documentation; a third is needed for health and safety and quality control; and a fourth is needed for waste management and equipment decontamination. If a raft or boat is used to assist the sampling procedure, at least one additional person will be needed.

The following equipment and procedure can be used to collect surface water samples for chemical and/or radiological analysis:

1. extendable bottle sampler
2. sample bottles
3. sample preservatives

   4. pH, temperature, and conductivity meters
   5. sample labels
   6. cooler packed with Blue Ice®
   7. trip blank and coolant blank
   8. sample logbook
   9. chain-of-custody forms
  10. chain-of-custody seals
  11. permanent ink marker
  12. health and safety screening instruments
  13. health and safety clothing
  14. waders, raft, or boat
  15. DOT-approved 55-gal waste drum
  16. sampling table
  17. plastic waste bags

Sampling Procedure

1. In preparation for sampling, read the introduction to Intrusive Sampling Methods (page 55) to confirm that all necessary preparatory work has been completed, including: obtaining property access agreements; meeting health and safety, decontamination, and waste disposal requirements; and calibrating all health and safety and sampling equipment.
2. Prior to sample collection, fill a clean glass jar with sample water and measure the pH, temperature, and conductivity of the water. Record this information in a sample logbook.
3. To collect a grab sample, lower the sampler to the desired sampling interval, and lift up on the bottle seal ring to allow the sample bottle to fill with water. When the bottle is full, depress the seal ring and retrieve the sample. Transfer water from the sampler into a sample bottle for laboratory analysis. A composite sample is collected by homogenizing several grab samples over either a horizontal or vertical sampling area (see Figure 3.30 and 3.31).
4. If the analyses to be performed require the sample to be preserved, this should be performed prior to filling the sample bottle.
5. After the bottle is capped, attach a sample label and custody seal to the bottle and immediately place it into a Blue Ice®-packed cooler. Samples to be analyzed for radionuclides do not commonly require cooling.
6. See Chapter 5 for details on preparing sample bottles and coolers for sample shipment.
7. Containerize any waste water in a DOT-approved 55-gal drum. Prior to leaving the site, all waste drums should be sealed, labeled, and handled appropriately (see Chapter 7).
8. Finally, the coordinates of the sampling point should be surveyed in by a professional surveyor to preserve the exact sampling location.

Extendable Tube Sampler Method. The Extendable Tube Sampler is an effective tool for collecting grab or composite samples as deep as 5 ft below the water surface. The sampler is composed of a sampling tube which contains a ball check-valve with plunger-type inner rod, and a vertical sampling pole with an inner rod to open and close the check valve (Figure 3.35).

This method and the Extendable Bottle Sampler Method (page 124) are particularly effective for characterizing streams and rivers which are suspected of having stratified zones of contamination, since they can collect grab samples from various depth intervals within a water column. While this method is most effective in collecting grab samples, it can also be used to collect composite samples by combining a number of grab samples together in one sample bottle (see Figures 3.30 and 3.31). A composite sample should not be collected when the analyses to be performed on the sample include volatile organics, since volatilization is facilitated by the compositing process. For a summary of the effectiveness and limitations of this sampling method, see Table 3.5.

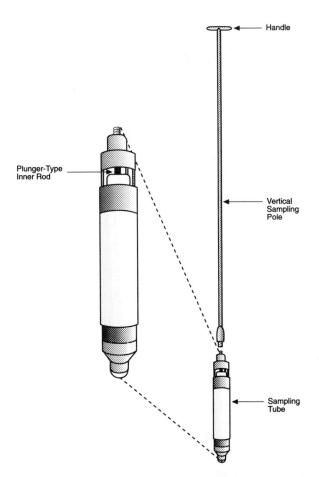

Figure 3.35.   Extendable Tube Sampler used to collect surface water samples.

For most sampling programs, four people are sufficient for this sampling procedure. Two are needed for field testing, sample collection, labeling, and documentation; a third is needed for health and safety and quality control; and a fourth is needed for waste management and equipment decontamination. If a raft or boat is used to assist the sampling procedure, at least one additional person will be needed.

The following equipment and procedure can be used to collect surface water samples for chemical and/or radiological analysis:

1. extendable tube sampler
2. sample bottles
3. sample preservatives
4. pH, temperature, and conductivity meters
5. sample labels
6. cooler packed with Blue Ice®
7. trip blank and coolant blank
8. sample logbook
9. chain-of-custody forms
10. chain-of-custody seals
11. permanent ink marker
12. health and safety screening instruments
13. health and safety clothing
14. waders, raft, or boat
15. DOT-approved 55-gal waste drum
16. sampling table
17. plastic waste bags

Sampling Procedure

1. In preparation for sampling, read the introduction to Intrusive Sampling Methods (page 55) to confirm that all necessary preparatory work has been completed, including: obtaining property access agreements; meeting health and safety, decontamination, and waste disposal requirements; and calibrating all health and safety and sampling equipment.
2. Prior to sample collection, fill a clean glass jar with sample water and measure the pH, temperature, and conductivity of the water. Record this information in a sample logbook.
3. To collect a grab sample, lower the sampler to the desired sampling interval, and open the ball check-valve to allow the sampling tube to fill with water. When the sampler is full, close the check-valve and retrieve the sample. The water is then transferred into a sample bottle through the bottom dump valve. A composite sample can be collected by combining a number of grab samples over either a vertical or horizontal sampling area.
4. If the analyses to be performed require the sample to be preserved, this should be performed prior to filling the sample bottle.
5. After the bottle is capped, attach a sample label and custody seal to the

bottle and immediately place it into a Blue Ice®-packed cooler. Samples to be analyzed for radionuclides do not commonly require cooling.

6. See Chapter 5 for details on preparing sample bottles and coolers for sample shipment.
7. Containerize any waste water in a DOT-approved 55-gal drum. Prior to leaving the site, all waste drums should be sealed, labeled, and handled appropriately (see Chapter 7).
8. Finally, the coordinates of the sampling point should be surveyed in by a professional surveyor to preserve the exact sampling location.

Pond, Lake, and Retention Basin Sampling

As part of an initial site characterization study, one surface water sample is commonly collected from each retention basin, and as many as two samples, one near the inlet and outlet, from each pond and/or lake. If contamination is identified in any of these samples, additional sampling should be considered to assist in defining the extent of contamination. The number of additional samples needed to complete the characterization is dependent on the size of the water body and the DQOs (see page 33).

The depth at which surface water samples are collected should be based on the suspected concentration and density of the contaminants of concern. If DNAPLs are suspected, samples should be collected from the base of the water column. Otherwise, samples are most commonly collected from the top of the water column. If historical information leads the investigator to believe that contamination could be layered, due to varying specific gravities of the contaminants, one should consider collecting either a vertical composite sample, or several grab samples from different depth intervals (Figure 3.31).

Although a number of sophisticated sampling devices are available to collect surface water samples, not all of these tools are effective in collecting samples from ponds, lakes, and retention basins. The methods that have proved to be the most effective when sampling these environments include the Bottle Submersion Method, Dipper Method, Extendable Bottle Sampler Method, Extendable Tube Sampler Method, Bailer Method, Kemmerer Bottle Method, and Bomb Sampler Method. The first four of these methods are identical to those used to collect water samples from streams, rivers, and surface water drainages (see pages 120–129), with a few minor modifications.

All seven of these methods are effective in collecting grab samples. Of these methods, the Bottle Submersion Method is the easiest to implement since it simply involves submersing a sample bottle just beneath the water surface. A telescoping extension rod is often used with this method to hold the sample bottle as it is lowered into the water. The Dipper Method is also very easy to implement since it simply involves lowering a dipper below the water surface, then transferring the water into sample bottles. These two methods are only effective in collecting samples from the water surface.

The Extendable Bottle Sampler Method utilizes a glass sample bottle attached to a vertical sampling pole. After lowering the bottle to the desired sampling depth, the seal ring is lifted, which allows the bottle to fill with water. The sampler is then retrieved and water is transferred into a sample bottle. This method is effective in collecting samples as deep as 5 ft below the water surface.

The Extendable Tube Sampler utilizes a sampling tube and a vertical extension rod. Similar to the Extendable Bottle Sampler, this method utilizes a check valve to open and close the sampling tube when the desired sampling depth is reached. After the sampling tube is retrieved, water is transferred into a sample bottle. This method is effective in collecting samples as deep as 5 ft below the water surface.

The bailer is composed of a sampling tube, pouring spout, and a bottom check valve. The bailer is lowered by hand just below the water surface. When the bailer is full it is retrieved, and the water is transferred into a sample bottle. This method collects a sample representative of the top several feet of the water column.

The Kemmerer Bottle is composed of a vertical sampling tube, center rod, trip head, and bottom plug. The bottle is lowered through the water by means of a sampling line to the desired sampling depth. The sampling tube is then opened with the assistance of a trip weight. When the sampler is full, the water is retrieved and transferred into a sample bottle. The depth at which a sample can be collected with this method is only restricted by the length of the sampling line.

The Bomb Sampler is composed of a sampling tube, center rod, and support ring. The sampler is lowered to the desired sampling depth using a support line. The center rod is then lifted using a second line which allows the sampler to fill with water. The water is then retrieved and transferred into a sample bottle. The depth at which a sample can be collected with this method is only restricted by the length of the sampling line.

Vertical composite samples can be collected using either the Extendable Bottle Sampler, Extendable Tube Sampler, Kemmerer Bottle, or Bomb Sampler. A vertical composite sample is collected by compositing several grab samples; each represents a different depth in the water column (Figure 3.31). Any of the seven methods can be used to collect areal composite samples, where grab samples from different locations are composited together (Figure 3.30). Any of the seven methods can also be used to collect an integrated sample where a portion of a sample is collected from the same location several times over a selected time period.

Table 3.6 summarizes the effectiveness of each of the seven sampling methods. A number "1" in the table indicates that a particular procedure is most effective in collecting samples for a particular laboratory analysis, sample type, or sampling depth. A number "2" indicates that the procedure is acceptable but less effective, while an empty cell indicates that the procedure is not recommended. For example, Table 3.6 indicates that Bottle Submersion is the preferred method for collecting samples for volatile organic analysis. While

Table 3.6. Evaluation Table for Pond, Lake, and Retention Pond Surface Water Sampling Methods

	Laboratory Analyses							Sample Type				Sampling Depth		
	Volatiles	Semi-Volatiles	Primary Metals	Pesticides	PCBs	TPH	Radionuclides	Grab	Composite (Vertical)	Composite (Areal)	Integrated	Surface (0.0-0.5 ft.)	Shallow (0.0-5.0 ft.)	Deep (>5.0 ft.)
Bottle Submersion	1	1	1	1	1	1	1	1		1	1	1		
Dipper	2	1	1	1	1	1	1	1		1	1	1		
Extendable Bottle Sampler	2	1	1	1	1	1	1	1	1	1	1	2	1	
Extendable Tube Sampler	2	1	1	1	1	1	1	1	1	1	1	2	1	
Bailer	2	1	1	1	1	1	1	1		1	1	1	2*	
Kemmerer Bottle	2	1	1	1	1	1	1	1	1	1	1		2	1
Bomb Sampler	2	1	1	1	1	1	1	1	1	1	1		2	1

*Able to collect a sample to a depth equal to the length of the bailer.
1 = Preferred Method
2 = Acceptable Method
Empty Cell – Method Is Not Recommended

the Dipper, Bailer, Extendable Bottle Sampler, Extendable Tube Sampler, Kemmerer Bottle, and Bomb Sampler are acceptable methods for volatile organic analysis, they are less desirable since the water from the sampler must be poured into sample bottles which tends to aerate the sample.

Whichever sampling method is selected, sample bottles for volatile organic analysis should always be the first to be filled. Bottles for the remaining parameters should be filled consistent with their relative importance to the sampling program.

Bottle Submersion Method. The Bottle Submersion Method used to collect grab surface water samples from ponds, lakes, and retention basins is identical to the procedure described for stream, river, and water drainage sampling (page 120), with the following modifications:

- If contaminants are identified from the preliminary sampling, the Kemmerer Bottle and Bomb Sampler should be considered to more accurately define the distribution of contaminants in addition to the Extendable Bottle Sampler and Extendable Tube Sampler Methods.
- If a raft or boat is used to assist the sample collection, a fifth person will be needed on the sampling team. This person's responsibilities are to maneuver and steady the boat.
- Add "waders, raft, or boat" to the equipment list.
- Modify Step 4 to read "While standing on the edge of the water body, or leaning over the side of the boat, extend the rod out over the water and lower

the sample bottle just below the water surface. If the analyses to be performed require the sample to be preserved, this should be performed prior to filling the sample bottle."

Dipper Method. The Dipper Method used to collect grab surface water samples from ponds, lakes, and retention basins is identical to the procedure described for stream, river, and water drainage sampling (page 122), with the following modifications:

- If contaminants are identified from the preliminary sampling, the Kemmerer Bottle and Bomb Sampler should be considered to more accurately define the distribution of contaminants in addition to the Extendable Bottle Sampler and Extendable Tube Sampler Methods.
- If a raft or boat is used to assist the sample collection, a fifth person will be needed on the sampling team. This person's responsibilities are to maneuver and steady the boat.
- Add "waders, raft, or boat" to the equipment list.
- Modify Step 3 to read "While standing on the edge of the water body, or leaning over the side of the boat, lower the dipper just below the water surface, being careful not to disturb the underlying sediment."

Extendable Bottle Sampler Method. The Extendable Bottle Sampler Method used to collect grab surface water samples from ponds, lakes, and retention basins is identical to the procedure described for stream, river, and water drainage sampling (page 124), with the following modification:

- If contaminants are identified from the preliminary sampling, the Kemmerer Bottle and Bomb Sampler should be considered to more accurately define the distribution of contaminants.

Extendable Tube Sampler Method. The Extendable Tube Sampler Method used to collect grab or composite surface water samples from ponds, lakes, and retention basins is identical to the procedure described for stream, river, and water drainage sampling (page 127), with the following modification:

- If contaminants are identified from the preliminary sampling, the Kemmerer Bottle and Bomb Sampler should be considered to more accurately define the distribution of contaminants.

Bailer Method. A bailer is most commonly used to collect groundwater samples; however, it can also be used to collect surface water samples from ponds, lakes, and retention basins (EPA 1987). A standard bailer is composed of a bailer body which is available in various lengths and diameters, a pouring spout, and a bottom check valve which contains a check ball (Figure 3.36).

As the bailer is lowered into the water body, water flows into the bailer through the bottom check valve. When the sampler is retrieved, a check ball prevents water from escaping through the bottom of the sampler. The sample is then poured from the bailer through a spout into sample bottles. Bailers used to collect water samples for laboratory analysis should be made of Teflon

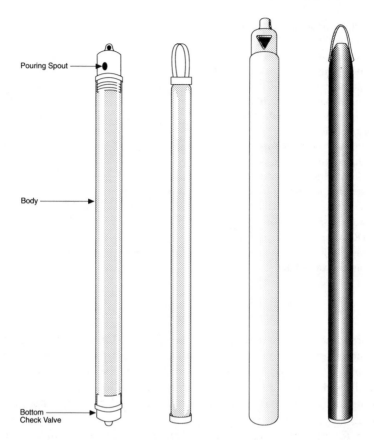

Figure 3.36. Various types of bailers available to collect surface water or groundwater samples.

or stainless steel. Depending on the depth of the water at the sampling point, samplers may need a pair of waders, raft, or boat to access the sampling point.

Some common modifications to the bailer include the use of extension couples to increase the length of the bailer, and a controlled flow bottom assembly. The bottom assembly allows the bailer to be emptied through the bottom of the sampler, which reduces the opportunity for volatilization to occur.

For most sampling programs, four people are sufficient for this sampling procedure. Two are needed for field testing, sample collection, labeling, and documentation; a third is needed for health and safety and quality control; and a fourth is needed for miscellaneous tasks such as managing waste water drums and equipment decontamination. If a raft or boat is used to assist the sampling procedure, at least one additional person will be needed.

The following equipment and procedure can be used to collect surface water samples for chemical and/or radiological analysis:

1. Teflon or stainless steel bailer
2. sample bottles
3. sample preservatives
4. pH, temperature, and conductivity meters
5. sample labels
6. cooler packed with Blue Ice®
7. trip blank and coolant blank
8. sample logbook
9. chain-of-custody forms
10. chain-of-custody seals
11. permanent ink marker
12. health and safety screening instruments
13. health and safety clothing
14. waders, raft, or boat
15. DOT-approved 55-gal waste drum
16. sampling table
17. plastic waste bags

Sampling Procedure

1. In preparation for sampling, read the introduction to Intrusive Sampling Methods (page 55) to confirm that all necessary preparatory work has been completed, including: obtaining property access agreements; meeting health and safety requirements, decontamination, and waste disposal requirements; and calibrating all health and safety and sampling equipment.
2. Prior to sample collection, fill a clean glass jar with sample water and measure the pH, temperature, and conductivity of the water. Record this information in a sample logbook.
3. When properly positioned over the sampling point, hold the bailer just above the pouring spout, and slowly lower it just deep enough to fill the bailer. Retrieve the sampler and carefully transfer the water into a sample bottle.
4. If the analyses to be performed require the sample to be preserved, this should be performed just prior to filling the sample bottle.
5. After the bottle is capped, attach a sample label and custody seal to the bottle and immediately place it into a Blue Ice®-packed cooler. Samples to be analyzed for radionuclides do not commonly require cooling.
6. See Chapter 5 for details on preparing sample bottles and coolers for sample shipment.
7. Containerize any waste water in a DOT-approved 55-gal drum. Prior to leaving the site, all waste drums should be sealed, labeled, and handled appropriately (see Chapter 7).
8. Finally, the coordinates of the sampling point should be surveyed in by a professional surveyor to preserve the exact sampling location.

Kemmerer Bottle Method. The Kemmerer Bottle Sampler Method is effective for collecting at-depth grab samples of water from ponds, lakes, and retention basins. The sampler is composed of a vertical sampling tube, center rod, head plug, and bottom plug (Figure 3.37). A line attached to the top of the sampler is used to lower the sampler to the desired sampling depth. The head plug and bottom plug are then tripped open by sliding a trip weight down the sampling line. When the sampling tube is full of water, the sampler is retrieved, and the water is transferred into a sample bottle.

The sampling line should be monofilament, such as common fishing line, and should be discarded between sampling points. The line should be cut to a length long enough to reach the desired sampling depth, and it must be strong enough to lift the weight of the Kemmerer Bottle when it is full of water. Prior to using the sampling line, it should first be decontaminated in the same manner as other sampling equipment (see Chapter 4).

The effective sampling depth of this sampler is only limited by the length of the sampling line. Since this method is used primarily to collect deep water samples, a raft or boat is commonly required to assist the sampling procedure.

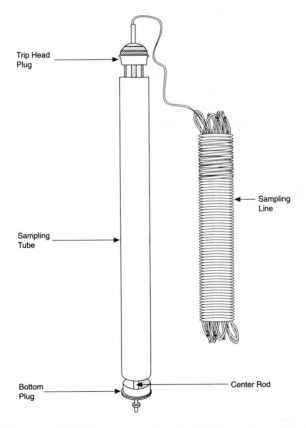

Figure 3.37. Kemmerer Bottle Sampler used to collect surface water samples.

When samples are to be collected from depths of 5 ft or less, the Extendable Bottle Sampler (page 132), or Extendable Tube Sampler (page 132) are recommended over the Kemmerer Bottle, since they are easier to use.

This method is particularly effective for characterizing ponds, lakes, and retention basins which are suspected to be contaminated with DNAPLs, since these contaminants tend to concentrate at the bottom of a water column. When DNAPLs and/or LNAPLs are present in a water body, it is not uncommon to have stratified zones of contamination. To characterize stratified conditions, either the Kemmerer Bottle or Bomb Sampler (pages 135 or 137) can be used to collect water samples from discrete intervals throughout the water column.

The Kemmerer Bottle is effective in collecting depth and areal composite water samples as well as integrated samples. A depth composite sample is acquired by compositing several grab samples, each representing a different depth in the water column. An areal composite sample is collected by compositing water samples from different locations, while an integrated sample is acquired by collecting a portion of a sample from the same location several times over an extended period of time.

For most sampling programs, five people are sufficient for this sampling procedure. Two are needed for field testing, sample collection, labeling, and documentation; a third is needed for health and safety and quality control; a fourth is needed for waste management and equipment decontamination; and a fifth is needed to operate the boat or maneuver the raft.

The following equipment and procedure can be used to collect surface water samples for chemical and/or radiological analysis:

1. Teflon or stainless steel Kemmerer Bottle
2. sampling line
3. sample bottles
4. sample preservatives
5. pH, temperature, and conductivity meters
6. sample labels
7. cooler packed with Blue Ice®
8. trip blank and coolant blank
9. sample logbook
10. chain-of-custody forms
11. chain-of-custody seals
12. permanent ink marker
13. health and safety screening instruments
14. health and safety clothing
15. waders, raft, or boat
16. DOT-approved 55-gal waste drum
17. sampling table
18. plastic waste bags

Sampling Procedure

1. In preparation for sampling, read the introduction to Intrusive Sampling Methods (page 55) to confirm that all necessary preparatory work has been completed, including: obtaining property access agreements; meeting health and safety, decontamination, and waste disposal requirements; and calibrating all health and safety and sampling equipment.
2. Prior to sample collection, fill a clean glass jar with sample water from the desired sampling depth, and measure the pH, temperature, and conductivity of the water. Record this information in a sample logbook.
3. When properly positioned over the sampling point, slowly lower the Kemmerer Bottle to the desired sampling depth.
4. Slide the trip weight down the sampling line to trip open the sample tube. When the sampler is full of water, retrieve it, and transfer the water into a sample bottle.
5. If the analyses to be performed require the sample to be preserved, this should be performed prior to filling the sample bottle.
6. After the bottle is capped, attach a sample label and custody seal to the bottle and immediately place it into a Blue Ice®-packed cooler. Samples to be analyzed for radionuclides do not commonly require cooling.
7. See Chapter 5 for details on preparing sample bottles and coolers for sample shipment.
8. Containerize any waste water in a DOT-approved 55-gal drum. Prior to leaving the site, all waste drums should be sealed, labeled, and handled appropriately (see Chapter 7).
9. Finally, the coordinates of the sampling point should be surveyed in by a professional surveyor to preserve the exact sampling location.

Bomb Sampler Method. The Bomb Sampler Method is effective for collecting at-depth grab samples of water from ponds, lakes, and retention basins. The sampler is composed of a sampling tube, center rod, and support ring (Figure 3.38). A line attached to the support ring is used to lower the sampler to the desired sampling depth. A second line is attached to the top of the spring-loaded center rod, and is used to open and close the sampling tube.

The support and sampling line should be monofilament, such as common fishing line, and should be discarded between sampling points. The line should be cut to a length long enough to reach the desired sampling depth, and must be strong enough to lift the weight of the Bomb Sampler when it is full of water. Prior to using the sampling line, it should first be decontaminated in the same manner as other sampling equipment (see Chapter 4).

After lowering the sampler to the desired sampling depth, the sampling line is lifted to allow the sampling tube to fill with water. When the sampling line is released, the center rod drops to reseal the sampling tube. The sampler is then retrieved and water is transferred into a sample bottle by placing the bottle

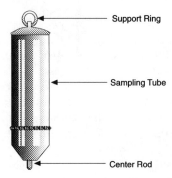

Figure 3.38.   Bomb Sampler used to collect surface water or groundwater samples.

beneath the center rod, and lifting up on the sampling line. Bomb samplers used to collect water samples for chemical or radiological analysis should be made of Teflon or stainless steel.

The effective sampling depth of this sampler is only limited by the length of the sampling line. Since this method is used primarily to collect deep water samples, a raft or boat is commonly required to assist the sampling procedure. When samples are to be collected from depths of 5 ft or less, the Extendable Bottle Sampler, or Extendable Tube Sampler are recommended over the Bomb Sampler, since they are easier to use.

The Bomb Sampler is effective in collecting deep grab and areal composite water samples as well as integrated samples. A depth composite sample is acquired by compositing several grab samples, each representing a different depth in the water column. An areal composite sample is collected by compositing water samples from different locations, while an integrated sample is acquired by collecting a portion of a sample from the same location, several times over a selected time period.

For most sampling programs, five people are sufficient for this sampling procedure. Two are needed for field testing, sample collection, labeling, and documentation; a third is needed for health and safety and quality control; a fourth is needed for waste management and equipment decontamination; and a fifth is needed to operate the boat or maneuver the raft.

The following equipment and procedure can be used to collect surface water samples for chemical and/or radiological analysis:

1. Teflon or stainless steel Bomb Sampler
2. sampling line
3. sample bottles
4. sample preservatives

 5. pH, temperature, and conductivity meters
 6. sample labels
 7. cooler packed with Blue Ice®
 8. trip blank and coolant blank
 9. sample logbook
10. chain-of-custody forms
11. chain-of-custody seals
12. permanent ink marker
13. health and safety screening instruments
14. health and safety clothing
15. waders, raft, or boat
16. DOT-approved 55-gal waste drum
17. sampling table
18. plastic waste bags

Sampling Procedure

1. In preparation for sampling, read the introduction to Intrusive Sampling Methods (page 55 to confirm that all necessary preparatory work has been completed, including: obtaining property access agreements; meeting health and safety, decontamination, and waste disposal requirements; and calibrating all health and safety and sampling equipment.
2. Prior to sample collection, fill a clean glass jar with sample water from the desired sampling depth, and measure the pH, temperature, and conductivity of the water. Record this information in a sample logbook.
3. When properly positioned over the sampling point, slowly lower the Bomb Sampler to the desired sampling depth.
4. Lift up on the sampling line and allow the sampling tube to fill with water. When the sampler is full, release the sampling line to close the reseal on the sampler.
5. Retrieve the sampler, and transfer the water into a sample bottle.
6. If the analyses to be performed require the sample to be preserved, this should be performed prior to filling the sample bottle.
7. After the bottle is capped, attach a sample label and custody seal to the bottle and immediately place it into a Blue Ice®-packed cooler. Samples to be analyzed for radionuclides do not commonly require cooling.
8. See Chapter 5 for details on preparing sample bottles and coolers for sample shipment.
9. Containerize any waste water in a DOT-approved 55-gal drum. Prior to leaving the site, all waste drums should be sealed, labeled, and handled appropriately (see Chapter 7).
10. Finally, the coordinates of the sampling point should be surveyed in by a professional surveyor to preserve the exact sampling location.

Groundwater Sampling

The following section provides the reader with guidance on selecting the most appropriate groundwater sampling method for the site under investigation. The criteria used to select the most appropriate method include the analyses to be performed on the sample, the type of sample to be collected (grab, composite, or integrated), and the sampling depth. SOPs have been provided for each of the recommended sampling methods to facilitate implementation. The following groundwater characterization strategies are provided as a supplement to the DQO process outlined in Chapter 2.

Whenever contamination is identified in soil, there is always the possibility of contaminants migrating to groundwater. This migration is possible through the transport mechanism of water percolating through the soil, while the rate of migration is controlled by soil physical properties such as pore size, and geochemical properties such as Distribution Coefficient (Kd), and organic carbon content. Once contaminants reach the groundwater they commonly disperse into the saturated formation. Depending on their physical and/or chemical properties, contaminants can concentrate near the top or bottom of the aquifer, or may evenly distribute themselves throughout the aquifer. Light Non-Aqueous Phase Liquids (LNAPLs), such as benzene, toluene, and xylene, have specific gravities less than water and therefore concentrate near the top of the aquifer. DNAPLs, such as PCE, TCE, and vinyl chloride, have specific gravities greater than water and therefore concentrate near the base of the aquifer. When dissolved, metals, VOCs, or petroleum hydrocarbons tend to more evenly distribute themselves throughout an aquifer.

When assessing the groundwater conditions at a site, serious consideration should be given to using the Direct Push Method (DPM) in combination with a mobile laboratory to perform a preliminary groundwater assessment prior to installing monitoring wells (see page 142). The advantages of the DPM are that groundwater samples can be collected quickly and inexpensively when compared to collecting the same data through monitoring well installation and sampling. Since a mobile laboratory can typically provide analytical results within hours after sampling, the DPM facilitates the "observational approach," where the results from samples analyzed in the field are used to guide the characterization effort. Once preliminary groundwater characterization is complete, groundwater wells can be precisely positioned for long-term monitoring, aquifer testing, and/or remediation.

If for some reason the DPM is not selected to perform preliminary groundwater characterization, one groundwater monitoring well should be installed downgradient from each suspected source of contamination, and one well upgradient (Figure 3.39). These wells should be built to screen the first encountered water-bearing unit. When the analytical results are obtained from samples collected from these wells, any contaminants identified in the downgradient samples which are not also found in the upgradient sample are contaminants derived from the site.

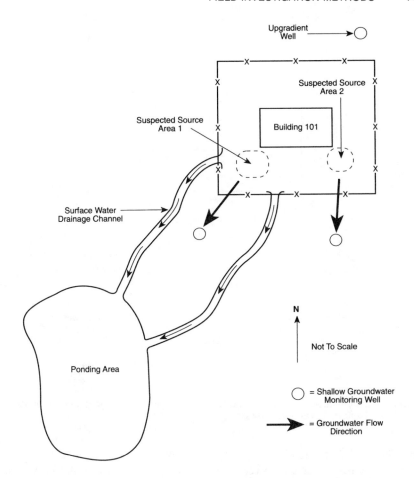

Figure 3.39. Common initial monitoring well configuration when the general groundwater flow direction is known.

Although shallow groundwater flow contours often mimic topographic contours, this is not always the case. When groundwater flow direction at a site is unknown, a minimum of three monitoring wells or piezometers should be positioned to form a triangle around the site, one of which should be positioned topographically downslope from the largest suspected source of contamination (Figure 3.40). This arrangement allows for an accurate determination of groundwater flow direction and gradient. After groundwater flow direction has been determined, one well should be installed downgradient from each suspected contaminant source. The wells should then be developed, purged and sampled. These samples should be analyzed for the contaminants of concern.

If groundwater contamination is identified in the upper aquifer, it is important to define the source, and lateral and vertical extent of contaminant migra-

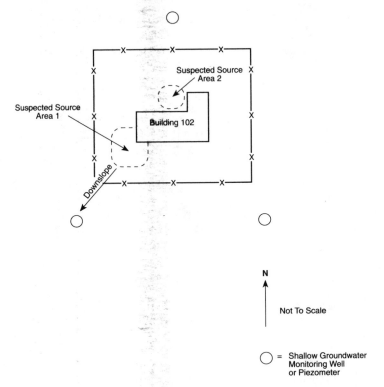

Figure 3.40. Common initial monitoring well configuration when the general groundwater flow direction is unknown.

tion. This is most effectively accomplished by using the DPM to thoroughly characterize the upper and lower aquifer, followed by the installation of long-term monitoring wells. The following section (Direct Push Method) provides an example of how this type of investigation would be performed.

Direct Push Method

The DPM utilizes a hydraulic press and slide hammer mounted on the rear end of a truck, to advance a sampling probe to a depth where a groundwater sample can be collected (Figure 3.41). Three sizes of trucks are available to perform this procedure. The size of the truck required is dependent on the depth of groundwater at the site. A lightweight van or truck is generally able to provide a reaction weight of 1,000- to 2,000-lb, which can advance a sampling probe 10- to 20-ft below the ground surface. The next larger truck provides a reaction weight of 3,000- to 5,000-lb, and is designed to sample as deep as 50 ft. The heaviest vehicle is the cone penetrometer (CPT) which has a reaction weight of 10- to 30-tons, and has been successful at penetrating through as much as 200 ft of unconsolidated sediment.

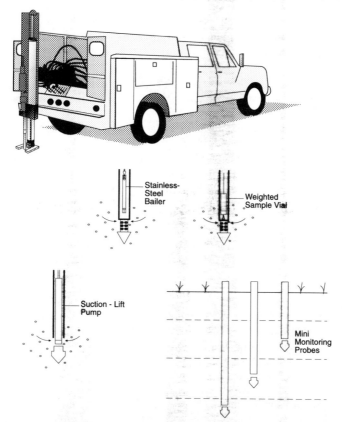

Figure 3.41. Direct Push groundwater sampling method.

The primary advantages of using DPM to assist a groundwater characterization study include:

- Very little investigation-derived waste is generated with the procedure
- Samples can be collected quickly
- The procedure is much less expensive than collecting the same data by installing and sampling monitoring wells
- The equipment gathers less public attention than a drill rig
- The procedure produces little disturbance to the surrounding environment.

Until recently, groundwater characterization has been performed by drilling and installing numerous groundwater monitoring wells in and around areas suspected of being contaminated. Since drilling procedures generate larger volumes of waste soil which must be drummed, stored, and ultimately disposed of, the less drilling that is required, the better. Although the need to install groundwater wells has not gone away, DPM can assist a groundwater investigation by selecting the optimum location for fewer wells.

The primary limitations of the DPM include the sampling depth, volume of sample that can be retrieved, and difficulties in penetrating through soils which contain gravel. Groundwater samples can be collected one of five different ways when using the DPM, including:

- Lowering a <0.5-in. diameter bailer down the inside of the probe multiple times to collect the volume of sample needed. This technique works well when only small volumes of water are needed.
- Lowering a weighted sample vial under vacuum down the inside of the probe. A needle inside the probe punctures the septum and allows water to flow into the vial. This method works well for collecting small volumes of water.
- A third method utilizes chambers in the probe which can be filled at depth, then brought to the surface. In most cases, the capacity of the chambers does not exceed 500-mL. To obtain larger volumes of water with this technique, the sampler must be advanced and retrieved repeatedly.
- Lowering a sample tube down the inside of the probe, and using a suction-lift pump to extract as much water as needed. This is the least preferred of all the available methods since it is not effective in collecting samples deeper than 25 ft, and is reported by the EPA to cause the volatilization of the sample, and possibly affect the pH (see page 171).
- A fifth method can be used to collect samples from formations with very low permeability. This technique involves running a screened tube down the inside of the sampling hole, removing the steel rods and packing sand around the screened section. This mini-monitoring well can be left in place as long as is required for water to fill the hole. Tests have shown that analytical results from water samples collected from the temporary wells compare favorably with data from conventional wells, while the temporary wells are a fraction of the cost. In addition to installing mini-monitoring wells, the direct push method can also be used to install temporary piezometers for water level monitoring.

If groundwater contaminants at a site include both LNAPLs and DNAPLs, it is recommended that the groundwater samples be collected from both the top and bottom of the aquifer, since this is where these contaminants will concentrate. To more completely characterize the distribution of contaminants, samples can be collected at regular intervals throughout the depth of an aquifer. This type of depth interval sampling is commonly performed at locations close to the suspected contaminant source(s), and the results are used to determine the most appropriate sampling interval for more distant sampling points.

If contamination is identified in the upper aquifer, the DPM can be used to collect samples from the next deeper aquifer, assuming it is within the depth penetration range of the sampler. The sampling interval(s) within the lower aquifer should be selected based on the types of contaminants identified in the upper aquifer.

To use the DPM most cost-effectively, groundwater investigations should be performed in combination with a mobile laboratory, and should be performed using the "observational approach." The "observation approach" utilizes the analytical data from each sampling point to decide where to position additional sampling points. This method is only possible when using a mobile laboratory that can provide analytical results shortly after sampling. This approach avoids the problem of collecting unnecessary or insufficient data.

An example of a successful groundwater investigation using the observational approach is illustrated in Figure 3.42. In this example, historical information led investigators to believe that buried tanks located south and west of Building 101 were potential sources of groundwater contamination at the site.

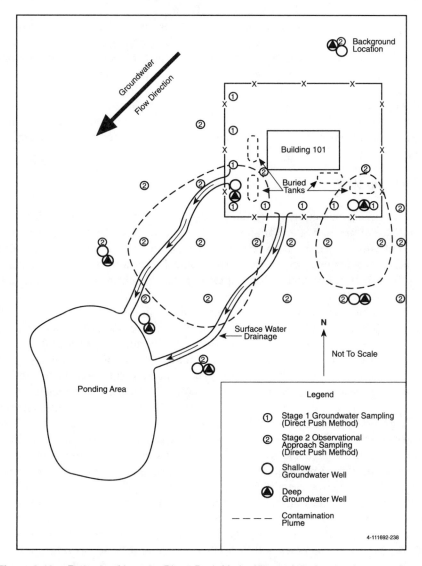

Figure 3.42. Example of how the Direct Push Method is used to characterize groundwater contamination using the observational approach.

The first step in this investigation involved collecting an initial row of groundwater samples in a "V" pattern, just inside the site property boundary, with the "V" pointing in the downgradient direction. Collecting initial groundwater samples from these locations will assure that any groundwater contamination leaving the site will be detected by one or more of these sampling points. Based on the results from this first phase of sampling, the observational approach is used to track the extent of the contaminant plume. This approach involves sampling outward from a contaminated sampling point in a grid pattern until the edge of the contaminant plume is defined.

Once the plume has been defined, monitoring wells should be installed for long-term monitoring purposes (Figure 3.42). These wells are commonly positioned near the source and downgradient from the leading edge of the contaminant plume(s) to track the long-term migration of the contamination. Shallow and deep well pairs are recommended to track both the vertical and horizontal migration of contaminants. The number of wells required is based on the size of the contaminant plume(s) identified. If groundwater remediation is later determined to be necessary, additional wells may be needed near the source of contamination.

Monitoring Wells

The primary objective of installing monitoring wells is to provide an access point where groundwater samples can be repeatedly collected, and groundwater elevations measured. When installing monitoring wells it is important to minimize the disturbance to the surrounding formation, and to construct a well from materials which will not interfere with the chemistry of the groundwater (EPA 1991).

The primary components of a groundwater monitoring well are the well screen, sump, riser pipe, well cap, protective steel casing, and lock (Figure 3.43). The well screen is by far the most critical component of a well. A well screen must have slots which are large enough to allow groundwater and contaminants to flow freely into a well, yet small enough to prevent formation soils from entering the well. The most common lengths of screen used to construct monitoring wells are 2-ft, 5-ft, 10-ft, 15-ft, and 20-ft. However, the EPA generally discourages the use of well screens greater than 10-ft since larger sampling intervals tend to dilute the sample. Another potential problem with using long screen lengths is that two aquifers can unintentionally be screened in the same well. Such an error provides a conduit for cross-contamination between aquifers.

When setting wells in an unconfined aquifer where LNAPLs may be present, the screen should be set so that it spans the vadose zone and the upper portion of the aquifer to allow the floating contaminants to enter the well (EPA 1991). For example, a 10-ft well screen should be set so that approximately 8-ft of the screen is below the mean static water level, and 2-ft is above. By constructing the well in this manner, any floating hydrocarbons that are

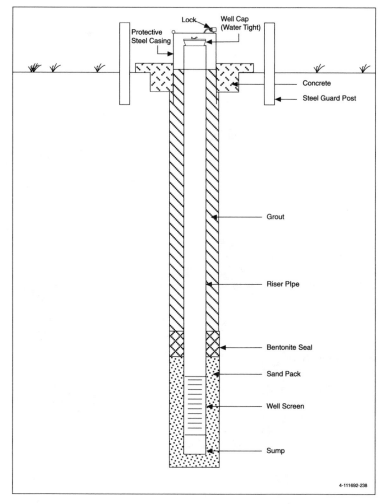

Figure 3.43.   Primary components of a groundwater monitoring well.

present in the formation will be visible in the well, even with seasonal fluctuations in groundwater levels. Similarly, when using a well screen that is 15-ft or 20-ft in length, 3-ft and 4-ft of the screen should be set above the mean static water level, respectively. Screen lengths of 2- and 5-ft are used to sample specific intervals within the aquifer and are not designed to screen the static water level.

When the contaminants of concern are DNAPLs, the bottom of the well screen should be set at the bottom of the aquifer (Figure 3.44). In a relatively thick aquifer where discrete layering of contamination is suspected to occur, several casings can be nested together in one large diameter well boring to allow the screening of several depth intervals (Figure 3.45).

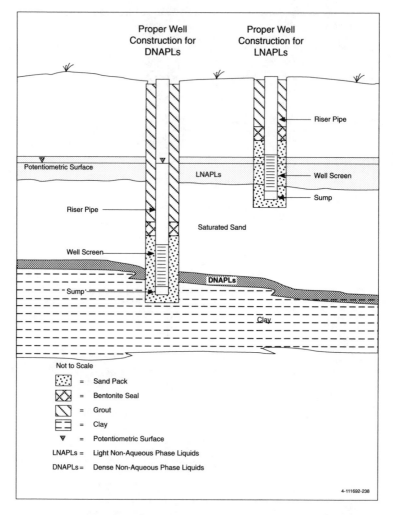

Figure 3.44. Common well construction in an unconfined aquifer when contaminants include DNAPLs and LNAPLs.

When setting a well in a confined aquifer where the contaminants of concern are expected to be dissolved in the water, or floating hydrocarbons, the top of the screened interval is positioned at the top of the aquifer (Figure 3.46). Since confined aquifers experience overburden pressures, the static water level often rises well above the top of the aquifer. Consequently, these wells are very seldom built to screen the static water level. If the contaminants of concern have specific gravities greater than water, the bottom of the well screen is set at the bottom of the confined aquifer.

Well screens for use in environmental sampling are available in stainless steel, Teflon, and polyvinyl chloride (PVC). Of these three materials, stainless

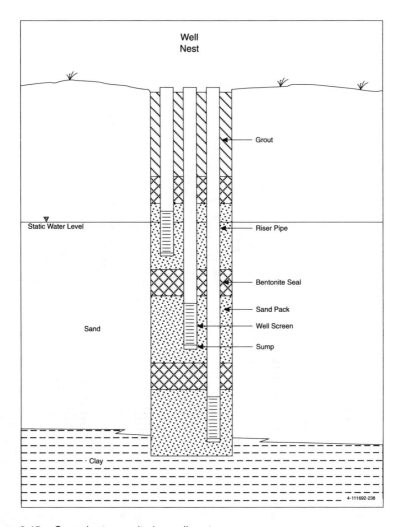

Figure 3.45. Groundwater monitoring well nest.

steel is preferred since it is relatively inert, durable, and of moderate cost. In contrast, Teflon is a very inert substance; however, it is very expensive, and since it is not as strong as stainless steel it can be easily damaged during well installation. PVC is relatively inexpensive; however, studies have shown that it can release and absorb trace amounts of various organic constituents after prolonged exposure. The most commonly used screen slot sizes range from 0.01-in. to 0.03-in. for silty to coarse sandy formations, respectively. However, screens can be special-ordered with slots as small as 0.006-in., and as large as needed. To determine the appropriate screen slot size for use in a particular formation, a sieve analysis must be performed either in the field or in a laboratory to determine the grain size distribution of the formation. With this

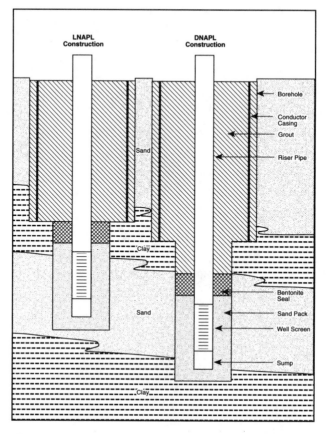

Figure 3.46. Common well construction in a confined aquifer when contaminants include DNAPLs and LNAPLs.

information, slot size calculations can be made using procedures outlined in Driscoll (1986).

A sump is threaded to the bottom of a well screen for the purpose of catching any fine grained soil which enters the well through the screen. If a sump is not used, soil accumulation will occur within the well screen and will eventually plug up the well. Common well sumps range in lengths from 0.5-ft to 3-ft, and should be made of the same material as the well casing. Sumps must be cleaned out on a regular basis to assure that the screen remains clear.

A threaded riser pipe is attached to the well screen to complete the well to the ground surface. The riser pipe is cut near the ground surface to allow for either an aboveground or belowground completion. For an aboveground completion, the riser pipe is typically cut 1.5-ft to 2.0-ft above the ground surface. A locking protective steel casing is then grouted in place over the riser pipe. A concrete pad and guard posts are often positioned around a well for added protection (Figure 3.43).

For a belowground completion, the riser pipe is cut several inches below the ground surface. A protective steel casing is then grouted in place over the riser pipe and completed flush with the ground surface (Figure 3.47). A watertight well cap should always be used in a belowground completion to prevent surface water from entering the well.

A sand pack is built around the well screen to a level 2-ft to 3-ft above the top of the screen. The purpose of the sand pack is to reduce the amount of fine grained formation soils from entering the well screen. The sand pack is built above the top of the screen to allow for the potential settling of the sand pack

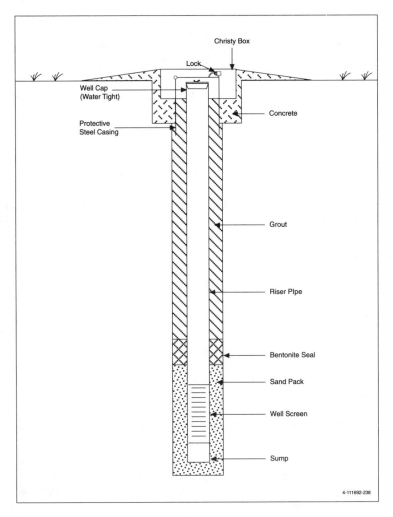

Figure 3.47.   Example of a belowground monitoring well completion.

over time. The grain size and size distribution of the sand pack used to build the well should be calculated using procedures outlined in Driscoll (1986), and should be compatible with the screen slot size. A 2-ft to 3-ft bentonite seal should be built over the sand pack using bentonite pellets. The remainder of the borehole should be filled with grout composed of portland type I/II cement mixed with approximately 4 to 5% bentonite powder.

Well Development. Prior to collecting a groundwater sample from a well, it must first be properly developed and then purged. When a well is installed using an auger rig, it is not uncommon for clayey soils to smear along the walls of the borehole. Similarly, when the mud rotary technique is used, a mudcake often develops along the walls of the borehole. The development procedure restores the aquifer's natural hydraulic conductivity and geochemical equilibrium near the well so that representative samples of formation water can be collected. In the development procedure, a well is surged and pumped long enough to clean out the well, and until the water stabilizes in pH, temperature, conductivity, turbidity, and clarity. Other measurements which should be considered when developing a well are dissolved oxygen, and oxidation reduction potential.

Surging is the first step in the development procedure which involves lowering a surge block down the well using a wire line from a drill rig or development truck. The surge block is repeatedly raised and lowered over 2-ft to 3-ft intervals starting at the top of the screened interval and moving downward (EPA 1992a). This procedure pulls soil particles that are finer than the well screen into the well so that it can later be removed. This procedure helps to compact the sand pack around the well screen, and should be performed several times during the development operation to assure that most of the fine grained soil particles have been removed from the immediate vicinity of the well screen. Following each surging step, a submersible pump and/or bailer are used to remove turbid water and sediment from the well.

A submersible pump is used to develop wells which can yield water at rates greater than several gallons per minute. In low yielding wells, a large bottom filling bailer is commonly used. When a submersible pump is used, the intake is first set near the bottom of the well screen, then near the center, and finally near the top. The pump is left at each interval until the water has cleared, and the pH, temperature, conductivity, and turbidity of the water have stabilized. For the most effective development, several different pumping rates are used at each development interval. The pumping rate for a submersible pump can be controlled by placing a restriction valve on the end of the discharge line, or by using a variable speed pump.

For poor producing aquifers, a bottom filling bailer is used for well development. Since the bailer method does not work as well as the submersible pump at pulling fine grained soil from the surrounding formation, the surging procedure is that much more important when developing these wells. After surging, the bailer is lowered to the bottom of the well using a wire line from a drill rig or development truck. When the bailer is full, it is retrieved to the ground surface, and the water is transferred into drums or a water holding tank. The bottom filling bailer not only removes groundwater from the well, but it also works effectively in removing any silt which has accumulated in the sump. In poor producing aquifers, it is tempting to want to add water to the well to assist the procedure; however, this should be avoided whenever possible because it can alter the groundwater chemistry. If a well will not develop without adding water, a field blank should be taken of the water added to the well, and it should be analyzed for the same parameters that the groundwater sample is to be analyzed for. Development methods using air should be avoided since they have the potential to alter the groundwater chemistry, and can damage the integrity of the well (EPA 1992a).

At regular intervals during the development procedure, a sample of the development water should be collected in a glass jar for pH, temperature, conductivity, and turbidity measurements. The results from these measurements are recorded on a Well Development Form (Chapter 5). A well is considered adequately developed when a minimum of three borehole volumes of water have been removed from the well, and the pH (± 0.1), temperature ($\pm 0.5°C$), conductivity ($\pm 10\%$), and clarity have stabilized, and turbidity measurements fall below 5-NTUs. A borehole volume is calculated using the formula (Figure 3.48):

$$V = \pi r^2 l$$

V = volume
π = 3.14
r = radius of the borehole
l = thickness of the water column

In clayey formations, it is not always possible to meet the < 5-NTU turbidity requirement. In this situation, development should continue until the turbidity of the water has stabilized. If the above physical parameters stabilize quickly, a minimum of 3 borehole volumes of water must be removed to consider the well adequately developed. In fine grained formations it is not uncommon for it to take 5 to 10 borehole volumes for all the physical parameters to stabilize.

Immediately after the completion of well development, the depth to groundwater should be measured using a water level probe, and recorded. This information can later be used to estimate the transmissivity of a formation.

If a well is built using the appropriate screen slot size and sand pack, very little siltation should occur over time. In this instance, future sampling only requires a well to be purged prior to sampling. However, if fine grained sediment begins building up inside a well sump and screen over time, the well will need to be redeveloped.

Well Purging. After development, a well should be allowed to set for several days, prior to purging and sampling. The well purging procedure is identical to

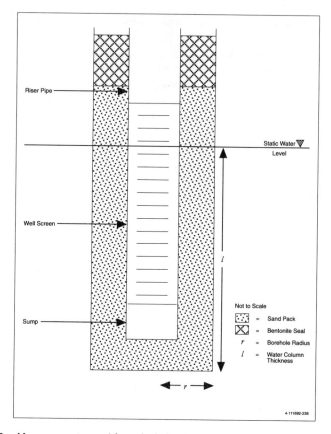

Figure 3.48. Measurements used for calculating borehole volumes for well development.

that for well development with the exception that surging is not performed. The objective of the purging procedure is to remove stagnant water from the well so that a representative water sample can be collected. The purging procedure should remove water throughout the screened interval to ensure that fresh formation water has replaced all stagnant water in the well (EPA 1992b). However, it is recommended that low pumping rates be used when purging to reduce the size of the cone of depression.

At regular intervals during the purging procedure, a sample of the purge water is collected in a glass jar for pH, temperature, conductivity, and turbidity measurements. The results from these measurements are recorded on a Well Purging Form (Chapter 5). Other measurements which should be considered when purging a well include dissolved oxygen, and oxidation reduction potential. In the purging procedure, water should be removed from a well until the pH (± 0.1), temperature ($\pm 0.5°C$), conductivity ($\pm 10\%$), turbidity (<5-NTUs), and clarity have stabilized.

It is important not to overpurge a well, since this may cause dilution or concentration of contaminants where stratification or leachates occur. High purge rates should also be avoided, since purging a well dry causes formation water to cascade into the well, allowing volatilization to occur. Purge rates should also never exceed development rates since this may draw new sediment into the well (EPA 1992b).

Immediately after the completion of well purging, the depth to groundwater should be measured using a water level probe, and recorded on the Well Purging Form. This information can later be used to estimate the transmissivity of the formation. As a general rule, one should attempt to sample a well before the static water level has had time to equilibrate. Wells which are slow to recover should be allowed to set only long enough for a sufficient sample volume of water to enter the well.

For very low yielding wells (<1-gpm), the above purging procedure is not practical since it would take days to complete. Rather, these wells should be bailed, or pumped dry twice, then sampled before groundwater equilibration.

Well Sampling. When collecting groundwater samples, bottles for volatile organic analysis should always be the first to be filled. This is critical since volatile organics are continuously being lost to the atmosphere during the sampling procedure. The remaining sample bottles should be filled in an order consistent with their relative importance to the sampling program. If the analyses to be performed require preservation, preservatives should be added to sample bottles prior to sample collection.

Shortly after a water sample has been collected from a well, it is common practice to take a final water level reading, and collect a final sample for pH, temperature, conductivity, and turbidity readings.

Collecting filtered and unfiltered groundwater samples for metals and/or radiological analysis should be considered, particularly for wells that remain cloudy or turbid throughout the development and purging procedure. Collecting samples in this way allows one to identify the dissolved concentration of metals and radionuclides. A pressurized filtration system utilizing 0.45 micron millipore membrane filter should be used when filtering samples.

The most effective sampling tools for collecting groundwater samples include the bailer, bomb sampler, bladder pump, piston pump, and submersible pump. The bailer is the least complicated of all the sampling tools and is commonly used to sample shallow wells. This tool is lowered down a well using a monofilament line. When the bailer is retrieved, the water is transferred into sample bottles. Similar to the bailer, the bomb sampler is lowered down the well using a monofilament line. This sampler is used to collect water or product samples from specific intervals in the water column. The bladder pump contains a Teflon bladder which fills and ejects water through a Teflon discharge line with the assistance of a compressed gas source such as a nitrogen bottle, or compressor. The piston pump is similar to the bladder pump in that it uses compressed gas as a power source; however, this pump utilizes a piston to force water out the discharge line. Submersible pumps are powered by an electric motor. This motor rotates impellers which force water up the discharge line.

Other groundwater sampling methods which are available but are not recommended include the Suction-Lift, and Air-Lift Methods. These sampling procedures have the tendency to cause the oxidation and degassing of the samples being collected. Other problems associated with these techniques are that the Suction-Lift Method has a tendency to affect the sample pH, and the Air-Lift Method can damage the integrity of the sand pack around the well screen if high pressures are used (EPA 1987, 1992a).

Table 3.7 summarizes the effectiveness of each of the four recommended procedures. A number "1" in the table indicates that a particular procedure is most effective in collecting samples for a particular laboratory analysis, sample type, or sample depth. A number "2" indicates that the procedure is acceptable but less effective, while an empty cell indicates that the procedure is not recommended. For example, Table 3.7 indicates that the Bailer, Bomb Sampler, Bladder Pump, and Piston Pump are all effective methods for collecting samples for volatile organic analysis, while the submersible pump method is an acceptable method for volatile organic analysis, but it is less effective than the other methods.

Whichever sampling method is selected, sample bottles for volatile organic analysis should always be the first to be filled. Bottles for the remaining parameters should be filled consistent with their relative importance to the sampling program.

Table 3.7. Evaluation Table for Groundwater Sampling Methods

	Laboratory Analyses							Sample Type			Sampling Depth	
	Volatiles	Semi-Volatiles	Primary Metals	Pesticides	PCBs	TPH	Radionuclides	Grab	Composite (Vertical)	Integrated	Shallow (0.0-30 ft.)	Deep (>30 ft.)
Bailer	2	1	1	1	1	1	1	1		2	1	2
Bomb Sampler	2	1	1	1	1	1	1	1	1	2	1	2
Bladder Pump	1	1	1	1	1	1	1	1	2	1	1	1
Piston Pump	2	1	1	1	1	1	1	1	2	1	1	1
Submersible Pump	2	1	1	1	1	1	1	1	2	1	1	1

1 = Preferred Method
2 = Acceptable Method
Empty Cell = Method is Not Recommended

1. Bailer Method

The Bailer Method is the simplest of all the groundwater sampling methods. A standard bailer is composed of a bailer body which is available in various lengths and diameters, a pouring spout, and a bottom check valve which contains a check ball (Figure 3.36). As the bailer is lowered into groundwater, water flows into the sampler through the bottom check valve. When the sampler is retrieved, the check ball seals the bottom of the bailer, which prevents water from escaping. Water is poured from the bailer through the pouring spout into sample bottles. Bailers used to collect water samples for chemical or radiological analysis should be made of Teflon or stainless steel.

Some common modifications to the bailer include the use of extension couples to increase the length of the bailer, and using a controlled flow bottom assembly. The bottom assembly allows the bailer to be emptied through the bottom of the sampler, which reduces the opportunity for volatilization to occur.

Some of the major advantages of the bailer are that it is easy to operate, portable, available in many sizes, and is relatively inexpensive. The disadvantages are that the sampling procedure is labor-intensive, and aeration of the sample can be a problem when transferring water into sample bottles.

The bailer should be lowered down a well using a monofilament line, such as common fishing line. The selected line should be cut to a length long enough to reach the groundwater, and it must be strong enough to lift the weight of the bailer when it is full of water. The line should be decontaminated in the same manner as the sampling bailer (see Chapter 4). The two most effective means of lowering a bailer and sampling line down a well are using the Hand-Over-Hand, or Tripod and Reel Methods.

To implement the Hand-Over-Hand Method, one end of the sampling line is tied to the top of the bailer, and the other end to the sampler's wrist. With the sampler's arms fully extended, the slack in the line is removed by wrapping it between the sampler's thumbs. To lower the bailer down the well, the sampler simply allows the line to unwind. To retrieve the bailer, the line is rewound. This method works very effectively for sampling wells which are less than 30 ft in depth, and when a small 2-in. diameter bailer is used.

To implement the Tripod and Reel Method, one end of the sampling line is tied to the top of the bailer, and the other end is tied to a reel (Figure 3.49). To lower the bailer down the well, the line is allowed to unwind from the reel. The handle on the reel is then used to rewind the line when the bailer is retrieved. This method is most effective in collecting samples from a depth less than 30 ft, and can handle a bailer as large as 4-in. in diameter.

For most sampling programs, four people are sufficient for the purging and sampling procedure. Two are needed for field testing, sample collection, labeling, and documentation; a third is needed for health and safety, and quality control; and a fourth is needed for miscellaneous tasks such as managing waste water drums, and equipment decontamination.

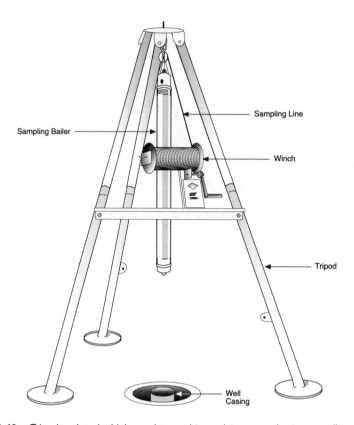

Figure 3.49. Tripod and reel which can be used to assist a groundwater sampling effort.

The following equipment and procedure can be used to collect groundwater samples for chemical and/or radiological analysis:

1. Teflon or stainless steel bailer
2. monofilament line
3. tripod and reel [not needed if the Hand-Over-Hand Method is selected]
4. water level probe
5. sample bottles
6. sample preservatives
7. pH, temperature, conductivity, and turbidity meters
8. wide-mouth glass jar
9. sample labels
10. cooler packed with Blue Ice®
11. trip blank and coolant blank
12. sample logbook
13. chain-of-custody forms
14. chain-of-custody seals
15. permanent ink marker
16. health and safety screening instruments

17. health and safety clothing
18. sampling table
19. DOT-approved 55-gal waste drum
20. plastic waste bags

Sampling Procedure

1. In preparation for sampling, read the introduction to Intrusive Sampling Methods (page 55) to confirm that all the necessary preparatory work has been completed, including: obtaining property access agreements; meeting health and safety, decontamination, and waste disposal requirements; and calibrating all health and safety, and sampling equipment.

2. Prior to sampling, a groundwater well must be properly developed and purged (see pages 152 and 153) If a well has been previously developed, there is no need to repeat this procedure unless accumulated sediment is blocking the well screen, or the well was inadequately developed the first time.

3. When well purging is complete, collect a final water level measurement, and record this information on the Well Purging and Sampling Form (see Chapter 5).

4. Cut a length of monofilament line long enough to reach the water table. Tie one end of the line to the top of the bailer, and the other end to either the sampling reel or sampler's wrist, depending on whether the Tripod and Reel or Hand-Over-Hand Method is used.

5. The bailer can be quickly lowered down the well to a depth just above the water table. At this point, slowly lower the bailer through the water just deep enough to fill it with water. When the bailer is full, slowly raise it out of the water, then retrieve it quickly to the ground surface. Collecting a groundwater sample in this manner creates little disturbance of the water column, which in turn reduces the loss of volatile organics, and minimizes the turbidity of the water sample.

6. Transfer the water from the bailer carefully into the appropriate sample bottle(s). If the analyses to be performed require the sample to be preserved, this should be performed prior to filling the sample bottle.

7. After the bottle is capped, attach a sample label and custody seal to the bottle and immediately place it into a Blue Ice®-packed cooler. Samples to be analyzed for radionuclides do not commonly require cooling.

8. See Chapter 5 for details on preparing sample bottles and coolers for sample shipment.

9. Fill a glass jar with sample water and collect and record a final pH, temperature, conductivity, and turbidity measurement, in addition to collecting a final water level measurement. Record this information on the Well Purging and Sampling Form.

10. Replace the well cap, lock the well, and containerize any waste in a DOT-

approved 55-gal drum. Prior to leaving the site, all waste drums should be sealed, labeled, and handled appropriately (see Chapter 7).

2. Bomb Sampler Method

The Bomb Sampler Method is similar to the Bailer Method, but provides the advantage of being able to collect a grab sample from a specific depth interval. A standard bomb sampler is composed of a sampling tube, center rod, and a support ring (Figure 3.38). With this method, a support line is used to lower the sampler to the desired sampling depth, while a sampling line is used to open and close the sampler inlet via the center rod. Within the body of the bomb sampler, a spring keeps the center rod in the closed position when lowering the sampler to the desired sampling depth, which prevents water from entering the sampler. When the desired sampling depth is reached, the sampling line is lifted against the pressure of the spring, which allows water to enter the sampling tube. When the sampling line is released, the center rod drops to reseal the sampling tube. After the sampler is retrieved, water is transferred into a sample bottle by placing the bottle beneath the center rod and lifting up on the sampling line. Bomb samplers used to collect water samples for chemical or radiological analysis should be made of Teflon and/or stainless steel.

Some of the major advantages of the bomb sampler are that it is effective for depth interval sampling, easy to operate, portable, available in many sizes, and is relatively inexpensive. The disadvantages are that the sampling procedure is relatively labor-intensive, and aeration of the sample can be a problem when transferring water into sample bottles.

The support and sampling line should be monofilament, such as common fishing line, and should be discarded between wells. The selected line should be cut to a length long enough to reach the desired sampling depth, and it must be strong enough to lift the weight of the sampler when it is full of water. The sampling line should be decontaminated in the same manner as other sampling equipment prior to sampling (see Chapter 4).

The most effective means of lowering a bomb sampler down a well is using the Tripod and Reel Method. To implement this method, one end of the sampling line is tied to the top of the sampler, and the other end is tied to the reel (similar to Figure 3.49). To lower the sampler down the well, the line is allowed to unwind from the reel. The handle on the reel is then used to rewind the line when retrieving the sampler.

For most sampling programs, four people are sufficient for the purging and sampling procedure. Two are needed for field testing, sample collection, labeling, and documentation; a third is needed for health and safety, and quality control; and a fourth is needed for miscellaneous tasks such as managing waste water drums, and equipment decontamination.

The following equipment and procedure can be used to collect groundwater samples for chemical and/or radiological analysis:

1. Teflon and/or stainless steel bomb sampler
2. monofilament line
3. tripod and reel
4. water level probe
5. sample bottles
6. sample preservatives
7. pH, temperature, conductivity, and turbidity meters
8. wide-mouth glass jar
9. sample labels
10. cooler packed with Blue Ice®
11. trip blank and coolant blank
12. sample logbook
13. chain-of-custody forms
14. chain-of-custody seals
15. permanent ink marker
16. health and safety screening instruments
17. health and safety clothing
18. sampling table
19. DOT-approved 55-gal waste drum
20. plastic waste bags

Sampling Procedure

1. In preparation for sampling, read the introduction to Intrusive Sampling Methods (page 55) to confirm that all the necessary preparatory work has been completed, including: obtaining property access agreements; meeting health and safety, decontamination, and waste disposal requirements; and calibrating all health and safety, and sampling equipment.
2. Prior to sampling, a groundwater well must be properly developed and purged (see pages 152 and 153). If a well has been previously developed, there is no need to repeat this procedure unless accumulated sediment is blocking the well screen, or the well was inadequately developed the first time.
3. When well purging is complete, collect a final water level measurement, and record this information on the Well Purging and Sampling Form (see Chapter 5).
4. Cut two lengths of monofilament line long enough to reach the sampling interval. Tie one end of one line to the top of the bomb sampler, and the other end to the sampling reel. The second line is tied to the top of the center rod.
5. The bomb sampler can be quickly lowered down the well to a depth just above the water table, then slowly lowered through the water to the sampling interval.
6. When the sampling interval is reached, hold the support line steady while lifting up on the sampling line. When the sampler is full of water, release the sampling line. Using the support line, slowly raise the sampler just

above the water surface, then retrieve it quickly to the ground surface. Collecting a groundwater sample in this manner creates little disturbance of the water column, which in turn reduces the loss of volatile organics, and minimizes the turbidity of the water sample.

7. Transfer water from the sampler carefully into sample bottles by placing the bottle beneath the center rod and lifting up on the center rod. If analyses to be performed require the sample to be preserved, this should be performed prior to filling the sample bottle.

8. After the bottle is capped, attach a sample label and custody seal and immediately place it into a Blue Ice®-packed cooler. Samples to be analyzed for radionuclides do not commonly require cooling.

9. See Chapter 5 for details on preparing sample bottles and coolers for sample shipment.

10. Fill a glass jar with sample water and collect and record a final pH, temperature, conductivity, and turbidity measurement, in addition to collecting a final water level measurement. Record this information on a Well Purging and Sampling Form.

11. Replace the well cap, lock the well, and containerize any wastewater in a DOT-approved 55-gal drum. Prior to leaving the site, all waste drums should be sealed, labeled, and handled appropriately (see Chapter 7).

3. Bladder Pump Method

A bladder pump combined with a source of compressed air, and an electronic controller, can be used as a very effective well purging and sampling tool. The bladder pump is composed of: a stainless steel pump body, bottom and top check ball, fill tube, Teflon bladder, outer sleeve, and discharge and air supply line (Figure 3.50). The advantages of this sampling tool are that it can be used effectively as either a dedicated or portable pump, it is a very clean system where the compressed air does not contact the sample water, it is effective at collecting samples as deep as 250 ft, and it is regarded highly by the EPA as an effective tool to collect samples for all parameters including VOCs (EPA 1992a). The disadvantages of this method are the relatively high equipment cost and low pumping rate.

This pump operates on an alternating fill and discharge cycle. During the fill cycle, the bottom check ball allows the bladder to fill with water, while the upper check ball prevents any liquid in the discharge line from dropping back into the pump. During the discharge cycle, the bottom check ball seats as compressed gas squeezes the bladder, which in turn forces water up the discharge line. Both the air supply and discharge lines should be made of Teflon.

The bladder pump is powered by a source of compressed air such as a gasoline-powered or electric-powered compressor, or a nitrogen gas cylinder. As a general rule, 0.5-psi is required per foot of well depth, plus 10-psi for pressure that is lost in the pressure line. If a gasoline-powered compressor is used, extreme care must be taken not to contaminate the water sample with

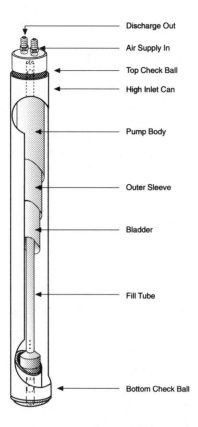

Discharge Out

Air Supply In

Top Check Ball

High Inlet Can

Pump Body

Outer Sleeve

Bladder

Fill Tube

Bottom Check Ball

Figure 3.50.   Bladder Pump used to collect groundwater samples.

gasoline or exhaust derived from the compressor. An electronic controller is used to control the timing of the bladder pump's fill and discharge cycle. The controller also monitors and controls the air pressure.

A 1.6-in. diameter bladder pump is able to pump water at a rate between 1.0- and 1.5-gpm at a depth of 25 ft. At a depth of 100 ft the pumping rate typically drops to around 0.5-gpm. These pumps are ideal for use in 2-in. diameter wells, but can also be used in larger diameter wells. Bladder pumps are also available in a 3.0-in. diameter size, which is ideal for larger diameter wells. This larger pump is able to pump water at a rate between 3- and 4-gpm at a depth of 25 ft, and 1- to 2-gpm at 100 ft.

When the wells at a site are to be sampled regularly for an extended period of time, it is recommended that one bladder pump be installed in each ground-water well. This is an expensive recommendation for any site; however, a significant amount of sampling time and money will be saved in the long run by not having to decontaminate the pump and discharge line between wells.

Also, equipment blank samples are not needed if a dedicated sampling system is used.

For most sampling programs, four people are sufficient for the purging and sampling procedure. Two are needed for field testing, sample collection, labeling, and documentation; a third is needed for health and safety, and quality control; and a fourth is needed for miscellaneous tasks such as managing waste water drums, and equipment decontamination.

The following procedure can be used to collect a groundwater sample for chemical and/or radiological analysis. The following equipment is needed for this sampling procedure:

1. bladder pump
2. Teflon discharge and air supply line
3. air compressor or compressed nitrogen gas bottle
4. electronic controller
5. water level probe
6. sample bottles
7. sample preservatives
8. pH, temperature, conductivity, and turbidity meters
9. wide-mouth glass jar
10. sample labels
11. cooler packed with Blue Ice®
12. trip blank and coolant blank (if applicable)
13. sample logbook
14. chain-of-custody forms
15. chain-of-custody seals
16. permanent ink marker
17. health and safety screening instruments
18. health and safety clothing
19. sampling table
20. DOT-approved 55-gal waste drum
21. plastic waste bags

Sampling Procedure

1. In preparation for sampling, read the introduction to Intrusive Sampling Methods (page 55) to confirm that all the necessary preparatory work has been completed, including: obtaining property access agreements; meeting health and safety, decontamination, and waste disposal requirements; and calibrating all health and safety, and sampling equipment.
2. Prior to sampling, a groundwater well must be properly developed and purged (see pages 152 and 153). If a well was previously developed as part of an earlier sampling effort, there is no need to repeat this procedure unless accumulated sediment is blocking the well screen, or the water is turbid. A well must be purged each time it is sampled.
3. When well purging is complete, leave the pump running and collect a final

water level measurement, and record this information on the Well Purging and Sampling Form (see Chapter 5).

4. Reposition the pump intake to the desired sampling depth and reduce the pumping rate so that a slow but steady flow of water flows from the discharge line.

5. Begin sampling by filling bottles for volatile organic analysis first. The remaining sample bottles should be filled in an order consistent with their relative importance to the sampling program.

6. If the analyses to be performed require the sample to be preserved, this should be performed prior to filling the sample bottle.

7. After a bottle is capped, attach a sample label and custody seal and immediately place it into a Blue Ice®-packed cooler. Samples to be analyzed for radionuclides do not commonly require cooling.

8. See Chapter 5 for details on preparing sample bottles and cooler for sample shipment.

9. Fill a glass jar with sample water and collect and record final pH, temperature, conductivity, and turbidity measurement, in addition to collecting a final water level measurement. Record this information on a Well Purging and Sampling Form.

10. Replace the well cap, lock the well, and containerize any wastewater in a DOT-approved drum. Prior to leaving the site, all waste drums should be sealed, labeled, and handled appropriately (see Chapter 7).

4. Piston Pump Method

The Piston Pump is composed of a stainless steel pump body, fill and empty check valve, fill tube, piston, and discharge and air supply line (Figure 3.51). The advantages of this sampling method are that the water sample does not come in contact with the compressed gas, the system is portable, is effective in collecting samples at depths as great as several hundred feet, and is considered by the EPA to be an effective pump for collecting samples for all parameters, including VOCs (EPA 1987). The disadvantages of this method are that only low pumping rates can be obtained, particulate material may damage or inactivate the pump unless the suction line is filtered, and the equipment cost is relatively high.

This pump is similar to the bladder pump in that it uses compressed gas in an alternating fill and discharge cycle. However, with this method a double-action piston is used to eject the water from the pump as opposed to a bladder. In the first action, water enters the pump cylinder on a suction stroke, and is then pushed out the discharge line on the pressure stroke. Check valves are used to allow water to enter the cylinder on the suction stroke, and leave the cylinder on the power stroke.

The piston pump is powered by a source of compressed air such as a gasoline-powered or electric-powered compressor, or a nitrogen gas cylinder. As a general rule, 0.5-psi is required per foot of well depth, plus 10-psi for

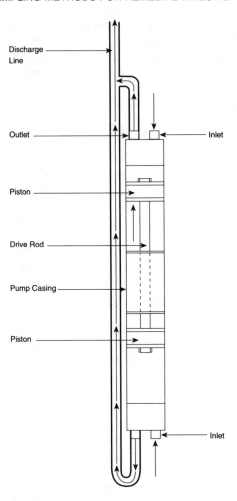

Discharge Line

Outlet

Inlet

Piston

Drive Rod

Pump Casing

Piston

Inlet

Figure 3.51. Piston Pump used to collect groundwater samples.

pressure that is lost in the pressure line. If a gasoline-powered compressor is used, extreme care must be taken not to contaminate the water sample with gasoline or exhaust derived from the compressor. An electronic controller is used to control the timing of the piston pump's fill and discharge cycle. The controller also monitors and controls the air pressure.

A 1.6-in. diameter piston pump is able to pump water at a rate of approximately 1.5-gpm at a depth of 25 ft, and 0.5-gpm at 100 ft. These pumps are ideal for use in 2-in. diameter wells, but can also be used in larger diameter wells. Piston pumps are also available in a 3.0-in. diameter size, which is ideal for larger diameter wells. This larger pump is able to pump water at a rate of approximately 4-gpm at a depth of 25 ft, and 2-gpm at 100 ft.

When the wells at a site are to be sampled regularly for an extended period of time, it is recommended that one piston pump be installed in each groundwater well. This is an expensive recommendation for any site; however, a significant amount of sampling time and money will be saved in the long run by not having to decontaminate the pump and discharge line between wells. Also, equipment blank samples will not need to be collected if a dedicated sampling system is used.

For most sampling programs, four people are sufficient for the purging and sampling procedure. Two are needed for field testing, sample collection, labeling, and documentation; a third is needed for health and safety, and quality control; and a fourth is needed for miscellaneous tasks such as managing waste water drums, and equipment decontamination.

The following procedure can be used to collect a groundwater sample for chemical and/or radiological analysis. The following equipment is needed for this sampling procedure:

1. stainless steel piston pump
2. Teflon discharge and air supply line
3. air compressor or compressed nitrogen gas bottle
4. electronic controller
5. water level probe
6. sample bottles
7. sample preservatives
8. pH, temperature, conductivity, and turbidity meters
9. wide-mouth glass jar
10. sample labels
11. cooler packed with Blue Ice®
12. trip blank and coolant blank (if applicable)
13. sample logbook
14. chain-of-custody forms
15. chain-of-custody seals
16. permanent ink marker
17. health and safety screening instruments
18. health and safety clothing
19. sampling table
20. DOT-approved 55-gal waste drum
21. plastic waste bags

Sampling Procedure

1. In preparation for sampling, read the introduction to Intrusive Sampling Methods (page 55) to confirm that all the necessary preparatory work has been completed, including: obtaining property access agreements; meeting health and safety, decontamination, and waste disposal requirements; and calibrating all health and safety, and sampling equipment.
2. Prior to sampling, a groundwater well must be properly developed and

purged (see pages 152 and 153). If a well was previously developed as part of an earlier sampling effort, there is no need to repeat this procedure unless accumulated sediment is blocking the well screen, or the water is turbid. A well must be purged each time a well is sampled.

3. When well purging is complete, leave the pump running and collect a final water level measurement, and record this information on the Well Purging and Sampling Form (see Chapter 5).
4. Reposition the pump intake to the desired sampling depth, and reduce the pumping rate so that a slow but steady flow of water flows from the discharge line.
5. Begin sampling by filling bottles for volatile organic analysis first. The remaining sample bottles should be filled in an order consistent with their relative importance to the sampling program.
6. If the analyses to be performed require the sample to be preserved, this should be performed prior to filling the sample bottle.
7. After the bottle is capped, attach a sample label and custody seal and immediately place it into a Blue Ice®-packed cooler. Samples to be analyzed for radionuclides do not commonly require cooling.
8. See Chapter 5 for details on preparing sample bottles and cooler for sample shipment.
9. Fill a glass jar with sample water and collect and record a final pH, temperature, conductivity, and turbidity measurement, in addition to collecting a final water level measurement. Record this information on a Well Purging and Sampling Form.
10. Replace the well cap, lock the well, and containerize any waste-water in a DOT-approved drum. Prior to leaving the site, all waste drums should be sealed, labeled, and handled in accordance with RCRA requirements (see page 7).

5. Submersible Pump Method

In the past, submersible pumps have been used primarily for well development and to collect groundwater samples from depths that exceed the limitations of other sampling methods. Currently, more sophisticated submersible pump systems are designed for purging and sampling. These systems utilize a pump made of all stainless steel and Teflon components, a Teflon discharge line, an electronic flow rate control system, and a power generator (Figure 3.52).

The submersible pump utilizes an electric motor which rotates a number of impellers, which in turn force water up the discharge line. These pumps are effective in collecting samples hundreds of feet in depth, and depending upon the size of pump, they are able to pump at rates as low as 100-mL/min to 100-gpm or more. The relatively light weight of the smaller pumps allows the flexibility of using the pump in a dedicated or nondedicated mode.

The advantage of the electronic flow rate control system is that the sampler

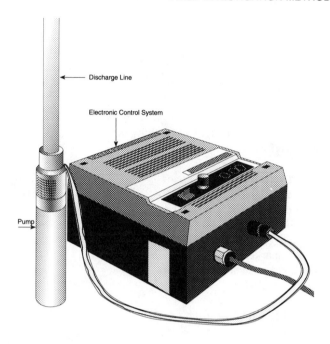

Discharge Line

Electronic Control System

Pump

Figure 3.52.   Submersible Pump used to collect groundwater samples.

can control the discharge rate of the pump without throttling the discharge line. Throttling is not desired when sampling, since it facilitates the loss of volatile organics. Other advantages include pumps being available in sizes small enough for use in 2-in. diameter wells, and the fact that EPA guidance considers this method suitable for collecting groundwater samples for all sampling parameters.

When the monitoring wells at a site are to be sampled regularly, it is recommended that dedicated pumps be installed in each well. This is an expensive recommendation for a large site; however, a significant amount of sampling time and money will be saved in the long run by not having to decontaminate the pump and discharge line between wells, nor will equipment blanks be needed each time the well is sampled.

For most sampling programs, four people are sufficient for the purging and sampling procedure. Two are needed for field testing, sample collection, labeling, and documentation; a third is needed for health and safety, and quality control; and a fourth is needed for miscellaneous tasks such as managing waste water drums, and equipment decontamination.

The following procedure can be used to collect a groundwater sample for chemical and/or radiological analysis. The following equipment is needed for this sampling procedure:

1. submersible pump
2. Teflon discharge line
3. generator or electrical power source
4. electronic flow rate controller
5. water level probe
6. sample bottles
7. sample preservatives
8. pH, temperature, conductivity, and turbidity meters
9. wide-mouth glass jar
10. sample labels
11. cooler packed with Blue Ice®
12. trip blank and coolant blank
13. sample logbook
14. chain-of-custody forms
15. chain-of-custody seals
16. permanent ink marker
17. health and safety screening instruments
18. health and safety clothing
19. sampling table
20. DOT-approved 55-gal waste drum
21. plastic waste bags

Sampling Procedure

1. In preparation for sampling, read the introduction to Intrusive Sampling Methods (page 55) to confirm that all the necessary preparatory work has been completed, including: obtaining property access agreements; meeting health and safety, decontamination, and waste disposal requirements; and calibrating all health and safety, and sampling equipment.
2. Prior to sampling, a groundwater well must be properly developed and purged (see pages 152 and 153). If a well was previously developed as part of an earlier sampling effort, there is no need to repeat this procedure unless accumulated sediment is blocking the well screen, or the water is turbid. A well must be purged each time a well is sampled.
3. When well purging is complete, leave the pump running and collect a final water level measurement, and record this information on the Well Purging and Sampling Form (see Chapter 5).
4. Reposition the pump intake to the desired sampling depth, and reduce the pumping rate so that a slow but steady flow of water flows from the discharge line.
5. Begin sampling by filling bottles for volatile organics first. The remaining sample bottles should be filled in an order consistent with their relative importance to the sampling program.
6. If the analyses to be performed require the sample to be preserved, this should be performed prior to filling the sample bottle.

7. After the bottle is capped, attach a sample label and custody seal and immediately place it into a Blue Ice®-packed cooler. Samples to be analyzed for radionuclides do not commonly require cooling.

8. See Chapter 5 for details on preparing sample bottles and cooler for sample shipment.

9. Fill a glass jar with sample water and collect and record a final pH, temperature, conductivity, and turbidity measurement, in addition to collecting a final water level measurement. Record this information on a Well Purging and Sampling Form.

10. Replace the well cap, lock the well, and containerize any waste-water in a DOT-approved drum. Prior to leaving the site, all waste drums should be sealed, labeled, and handled appropriately (see Chapter 7).

6. Suction-Lift Method

The Suction-Lift Method utilizes a surface-mounted pump, and is effective in collecting groundwater samples from wells where the water level is less than 25 ft below the ground surface. This system consists of a vacuum pump, collector flask, and sample collection tube. The method works by using the suction-lift pump to create a vacuum inside of the sampling tube. When the end of the sampling tube is positioned below the water table, air pressure on the water outside of the tube forces water to rise inside of the tube. As water rises to the top of the sampling line, it empties into the collector flask. The three most common types of suction-lift systems utilize a centrifugal, peristaltic, or diaphragm pump.

Some of the advantages that the Suction-Lift Method provides are that the equipment is relatively inexpensive, very portable, and it can be used in small diameter wells. The disadvantages of the method are that sampling is limited to shallow subsurface aquifers less than 25 ft in depth; pumping rates are generally low; degassing, loss of volatiles, and pH modification is possible (EPA 1987, 1992a).

Since the disadvantages of this method outweigh the advantages, the author does not recommend that this method be used. For shallow aquifers, the Bailer or Bladder Pump Method is easier to implement and provides higher quality samples (see pages 157 and 162). EPA guidance varies from do not use this method at all, to do not use the method to collect samples for volatile organic analysis (EPA 1987, 1992a).

7. Air-Lift Method

The Air-Lift Method utilizes a high-pressure hand pump or a small air compressor to force air into a well, which in turn forces water out of the discharge line. This method is not recommended since it introduces air into the sample which can cause oxidation, gas stripping, and other negative affects on the sample. This method can also damage the integrity of the filter pack around the well screen if high pressure evacuation is used (EPA 1987, 1992a).

Drum Sampling

The following section provides the reader with guidance and SOPs for collecting soil, sludge, and water samples from waste drums. In an effort to reduce the spread of contamination at a hazardous waste site, soil cuttings, decontamination water, well development and purging water, and other types of investigation-derived waste material should be drummed when appropriate and sampled to determine how to properly dispose of the material.

At all sites, samples should be analyzed using the TCLP method. This method measures the concentration of metals and organics that can be leached from a soil. If any of the TCLP guideline levels are exceeded, the material must be treated as a RCRA Hazardous Waste (see page 7 and Chapter 7). If TCLP guideline levels are not exceeded, one should determine whether any state or local requirements need to be met before releasing the material.

At radiological sites, samples should be analyzed for both TCLP and the radiological isotopes of concern. While the results from the TCLP analysis are compared to TCLP guidelines, the results from the radiological analyses are compared to background activity levels, as well as all applicable regulatory guidelines. If background and regulatory guidelines are not exceeded, one should next determine whether any state or local requirements need to be met before releasing the material.

Waste drums of unknown origin should be sampled to determine their content. Much greater health and safety concerns are related to the sampling of these drums since the contents may be toxic, corrosive, and/or explosive. It is common practice to pierce the lid of these types of drums with a spike by means of an unmanned mechanical device. By piercing the drum, the volatile organics in the headspace will be released, which in turn reduces the chances for explosion. The piercing device and other tools used to open these drums should be made from a "nonsparking" material. Using supplied air, chemical safety suits, and other health and safety precautions are essential when performing this type of sampling (see Chapter 6).

Soil Sampling from Drums

The methods recommended for collecting soil samples from drums for TCLP, and/or radiological testing are the Slide-Hammer and Hand Auger Methods. These methods are recommended over other soil sampling methods since they can be used effectively to collect composite soil samples throughout the depth of one or more drums.

Although a composite sample is not recommended when sampling soil specifically for volatile organic analysis, it is acceptable when sampling drums for TCLP analysis, since it provides more of a representative sample of the drum than simply taking a grab sample. It is generally acceptable to collect one composite sample from as many as three or four drums, preferably containing soil from the sample borehole. Compositing samples from more drums than this is not recommended, since it can dilute the sample beyond the point of providing reliable data.

Slide Hammer Drum Sampling Method. For collecting soil samples for drums for TCLP and/or radiological analysis, the Slide Hammer is very effective. This tool is comprised of a stainless steel core barrel, an extension rod, and a slide hammer (Figure 3.18). Most core barrels have an inside diameter of 2- or 2.5-in., and can be ordered in lengths as long as several feet. The top of the barrel is threaded so that it can be screwed into an extension rod. The hammer is available in different shapes and weights to accommodate the needs of the sampler.

For most sampling programs, four people are sufficient for this sampling procedure. Two are needed for sample collection, lithology description, labeling, and documentation; a third is needed for health and safety and quality control; and a fourth is needed for miscellaneous tasks such as waste management, and equipment decontamination.

The following equipment and procedure can be used to collect soil samples from drums for chemical and/or radiological analysis:

1. slide hammer and extension rod
2. stainless steel bowl
3. stainless steel spoon
4. sample jars
5. sample labels
6. cooler packed with Blue Ice®
7. trip blank and coolant blank
8. sample logbook
9. chain-of-custody forms
10. chain-of-custody seals
11. permanent ink marker
12. health and safety screening instruments
13. health and safety clothing
14. sampling table
15. nonsparking bung wrench
16. step pad
17. plastic waste bags

Sampling Procedure

1. In preparation for sampling, read the introduction to Intrusive Sampling Methods (page 55) to confirm that all necessary preparatory work has been completed, including: meeting health and safety, decontamination,

and waste disposal requirements; and calibrating all health and safety and sampling equipment.

2. Use a nonsparking wrench to remove the lid from the drum to be sampled.
3. While standing on a step pad, beat the sampling tube deep into the soil using the slide hammer.
4. Remove the sampler from the barrel by rocking it from side to side before lifting, or reverse beating the sampler from the hole.
5. Transfer the soil from the core barrel into a stainless steel bowl using a stainless steel spoon, and homogenize prior to filling a sample jar. If soil from more than one drum is being composited, transfer the soil from each drum into the same bowl and homogenize prior to filling a sample jar.
6. After the jar is capped, attach a sample label and custody seal and immediately place it into a Blue Ice®-packed cooler. Samples to be analyzed for radionuclides do not commonly require cooling.
7. See Chapter 5 for details on preparing samples and coolers for sample shipment.
8. Any soil left over from the sampling should be returned to the drum.
9. Finally, the drum should be resealed and handled appropriately (see Chapter 7).

Hand Auger Drum Sampling Method. For collecting soil samples for drums for TCLP and/or radiological analysis, the Slide Hammer is very effective. This tool is composed of a bucket auger, which comes in various shapes and sizes, a shaft, and a T-bar handle (Figure 3.17). Since the auger rotation automatically homogenizes the sampling interval, this method is very effective in collecting composite samples from waste drums.

For most sampling programs, four people are sufficient for this sampling procedure. Two are needed for sample collection, lithology description, labeling, and documentation; a third is needed for health and safety and quality control; and a fourth is needed for miscellaneous tasks such as waste management, and equipment decontamination.

The following equipment and procedure can be used to collect shallow soil samples for chemical and/or radiological analysis:

1. stainless steel hand auger
2. stainless steel bowl
3. stainless steel spoon
4. sample jars
5. sample labels
6. cooler packed with Blue Ice®
7. trip blank and coolant blank
8. sample logbook
9. chain-of-custody forms
10. chain-of-custody seals
11. permanent ink marker
12. health and safety screening instruments

13. health and safety clothing
14. sampling table
15. nonsparking bung wrench
16. step pad
17. plastic waste bags

Sampling Procedure

1. In preparation for sampling, read the introduction to Intrusive Sampling Methods (page 55) to confirm that all necessary preparatory work has been completed, including: meeting health and safety, decontamination, and waste disposal requirements; and calibrating all health and safety and sampling equipment.
2. Use a nonsparking wrench to remove the lid from the drum to be sampled.
3. While standing on a step pad, begin sampling by applying a downward pressure on the auger while rotating in a clockwise direction. When the auger is full of soil it should be removed from the hole, and the soil transferred into a stainless steel bowl using a stainless steel spoon. Continue sampling in this manner until the bottom of the drum is reached.
4. Composite the soil in the sampling bowl using the stainless steel spoon, then transfer into a sample jar. If soil from more than one drum is being composited, transfer the soil from each drum into the same bowl and homogenize prior to filling a sample jar.
5. After the jar is capped, attach a sample label and custody seal and immediately place it into a Blue Ice®-packed cooler. Samples to be analyzed for radionuclides do not commonly require cooling.
6. See Chapter 5 for details on preparing samples and coolers for sample shipment.
7. Any soil left over from the sampling should be returned to the drum.
8. Finally, the drum should be resealed and handled appropriately (see Chapter 7).

Sludge, Water, and Product Sampling from Drums

The method recommended for collecting sludge, water, or product samples from drums for analytical testing is the Coliwasa Method. This method is preferred over other methods because the sampling tube is thin enough to fit through a bung hole in the lid of a drum, and is capable of collecting either a grab sample from a specific depth within the drum, or can be used to collect a depth composite sample.

For drums containing investigation-derived wastewater or sludge, one depth composite sample is typically collected from each drum and analyzed for both the analytes on the TCLP list, to determine if it is a RCRA Hazardous Waste, and the specific contaminants of concern at that particular site. If the waste does not exceed any of the TCLP limits, it is not a RCRA Hazardous Waste.

However, prior to releasing the water or sludge to a storm sewer, it is also necessary to show that the water or sludge does not exceed drinking water MCLs. If an MCL is exceeded, the state will most likely require some form of treatment prior to release.

For drums containing waste chemicals or product, one depth composite sample is typically collected from each drum and analyzed for a suite of chemical analyses to determine its composition. If any of the components are compounds on the TCLP list, the material must be disposed of as a Hazardous Waste. Otherwise, one of the major manufacturers of the compound should be contacted to advise you on how to properly dispose of the waste. In some cases the waste can be recycled.

Coliwasa Method. The Coliwasa Method is effective for collecting grab and composite samples for sludge, water, or product samples from waste drums. This sampler is composed of a vertical sampling tube, piston suction plug, and handle (Figure 3.53). The inlet for the sampler is located at the base of the sample tube.

To collect a grab sample, the bottom of the sampler is lowered to the desired sampling depth. The suction plug is then raised to draw the sample into the sampling tube. When the sampler is full, it is removed from the drum and its contents are transferred into a sample bottle. On the other hand, to collect a composite sample, the bottom of the sampler is positioned at more than one sampling depth. At each depth, the suction plug is raised to fill a portion of the sampling tube. When the sampler is full, it is removed from the drum and its contents are transferred into a sample bottle.

For most sampling programs, four people are sufficient for this sampling procedure. Two are needed for field testing, sample collection, labeling, and documentation; a third is needed for health and safety, and quality control; and a fourth is needed for waste management and equipment decontamination.

The following equipment and procedure can be used to collect a grab or composite sludge, water, or product sample for chemical and/or radiological analysis:

1. Teflon Coliwasa Sampler
2. sample bottles
3. sample preservatives
4. pH, temperature, and conductivity meters
5. wide-mouth glass jar
6. sample labels
7. cooler packed with Blue Ice®
8. trip blank and coolant blank
9. sample logbook
10. chain-of-custody forms
11. chain-of-custody seals
12. permanent ink marker

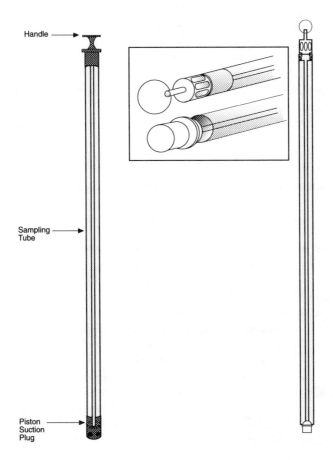

Handle

Sampling
Tube

Piston
Suction
Plug

Figure 3.53. Coliwasa Sampler used to collect water or sludge samples from waste drums.

13. health and safety screening instruments
14. health and safety clothing
15. sampling table
16. nonsparking bung wrench
17. step pad
18. unmanned mechanical spiking tool [only needed when sampling potentially explosive waste drums].

Sampling Procedure

1. In preparation for sampling, read the introduction to Intrusive Sampling Methods (page 55) to confirm that all necessary preparatory work has been completed, including: meeting health and safety, decontamination, and waste disposal requirements; and calibrating all health and safety and sampling equipment.

2. When sampling a waste drum where the contents are suspected to be highly toxic, corrosive, and/or explosive, or the contents are unknown, begin by piercing the lid of the drum with a "nonsparking" spike, preferably using an unmanned mechanical device. After allowing the headspace of the drum to aerate for several hours, loosen the ring bolt or bung using a nonsparking wrench, and remove the lid.

   When sampling drums containing investigation-derived waste, there is rarely a need to spike the drum prior to removing the lid. However, as a precaution, it is recommended that a nonsparking wrench be used to loosen the ring bolt or bung.

3. While standing on the step pad, lower the Coliwasa Sampler into the drum. If a grab sample is to be collected, position the inlet for the sampler at the desired sampling depth, then raise the piston suction plug to draw the sample into the sampling tube. When the sampler is full, remove it from the drum and transfer its contents into a sample bottle.

   To collect a composite sample, lower the bottom of the sampler to several sampling depths. At each depth, the suction plug is raised to fill a portion of the sampling tube. When the sampler is full, remove it from the drum and transfer its contents into a sample bottle.

4. After the bottle is capped, attach a sample label and custody seal and immediately place it into a Blue Ice®-packed cooler. Samples to be analyzed for radionuclides do not commonly require cooling.

5. See Chapter 5 for details on preparing samples and coolers for sample shipment.

6. Any sludge, water, or product left over from the sampling should be returned to the drum.

7. Finally, the drum should be resealed and handled appropriately (see Chapter 7).

REFERENCES

Acker, W.L. III, Basic Procedures for Soil Sampling and Core Drilling, Acker Drilling Company, Inc., Scranton, Pennsylvania, 56–57, 1974.

Byrnes, M.E., Complementary Investigative Techniques for Site Assessment with Low-Level Contaminants, Groundwater Monitoring Review, Fall, 90, 1990.

DeVera, E.R., B.P. Simmons, R.D. Stephens, and D.L. Storm, Samplers and Sampling Procedures for Hazardous Waste Streams, EPA 600/2-80-018, January 1980.

Driscoll, F.G., Groundwater and Wells, Johnson Division, St. Paul, Minnesota, 438-446, 1986.

Environmental Monitoring System Laboratory (EMSL), ORD, Environmental Protection Agency, Characterization of Hazardous Waste Sites – A Method Manual, Volume II – Available Sampling Methods, Las Vegas, Nevada, 1983.

Environmental Protection Agency, A Compendium of Superfund Field Operations Methods: Volume 1, EPA/540/P-87/001a, 1987.

Environmental Protection Agency, A Compendium of Superfund Field Operations Methods, 540/P-87/001A, Volume 2, 10-41 – 10-50, 10-61, 1987.

Environmental Protection Agency (Region VIII), Draft Standard Operation Procedures For Field Samplers, Rev. 4, 1992.

Environmental Protection Agency, Drum Handling Practices at Hazardous Waste Sites, EPA/600/S2-86/013, 1-4, 1986.

Environmental Protection Agency, Handbook for Sampling and Sample Preservation of Water and Wastewater, EPA/600/4-82-029, 1982.

Environmental Protection Agency, Handbook of Suggested Practices for the Design and Installation of Ground-Water Monitoring Wells, EPA/600/4-89/034, 1-123, 1991.

Environmental Protection Agency, Interim Methods for the Sampling and Analysis of Priority Pollutants in Sediment and Fish Tissue, Cincinnati, Ohio, ESML, October 1980b.

Environmental Protection Agency, RCRA Facility Investigation (RFI) Guidance Volume I, PB89-200299, 7-19 – 7-23, 1989.

Environmental Protection Agency, RCRA Facility Investigation (RFI) Guidance Volume II, PB89-200299, 9-66 – 9-69, 11-11, 1989.

Environmental Protection Agency, RCRA Facility Investigation (RFI) Guidance Volume II and III, PB89-200299, 9-64 – 9-78, 1989.

Environmental Protection Agency, RCRA Ground-Water Monitoring: Draft Technical Guidance, PB93-139350, 1992b.

Environmental Protection Agency, Soil Gas Monitoring Techniques Videotape, National Audio Visual Center, Capital Heights, Maryland, 1987.

Environmental Protection Agency, Soil Sampling Quality Assurance User's Guide, EPA/600/8-89/046, 1989.

Ford, P.J., et al., Characterization of Hazardous Waste Site – A Methods Manual, Volume II, Available Sampling Methods, NTIS PB85-168771, EPA 600/4-84-076, Las Vegas, Nevada, 1984.

Northeast Research Institute, Inc., Guide to Petrex Environmental Survey Field Procedures, 20, 1992.

Marrin, D.L. and G.M. Thompson, Remote Detection of Volatile Organic Contaminants in Ground Water via Shallow Soil Gas Sampling, Proceedings of the NWWA/API Conference on Petroleum Hydrocarbons and Organic Chemicals in Ground Water: Prevention, Detection and Restoration, National Water Well Association, Dublin, Ohio, 172-187, 1984.

Mason, B.J., Preparation of a Soil Sampling Protocol: Techniques and Strategies, NTIS PB83-206979, EPA, Las Vegas, Nevada, 1983.

Microseeps, Methods and Procedures for Use of Microseeps Soil Gas Sampling System, pp. 1-8, 1992.

Personal communication with representatives from Westinghouse Hanford, Washington, R.J. Electronics, Turner, Oregon, and RSI Research Ltd., Sidney, British Columbia, Canada.

Smith, R., and G.V. James, The Sampling of Bulk Materials, London, The Royal Society of Chemistry, 1981.

Target Environmental Services, Site Investigation and Remediation with "Direct Push Sampling Technology," Columbia, Maryland, 11, 1993.

Wyatt, D.E., Pirkle, R.J., and Masdea, D.J., Soil Gas Investigations at the Sanitary

Landfill (U), Westinghouse Savannah River Company, WSRC-RP-92-878, 1992, Appendix I.

BIBLIOGRAPHY

Anderson, M.P., Movement of Contaminants in Ground Water: Ground Water Transport-Advection and Dispersion; Ground-Water Contamination, Studies in Geophysics, National Academy Press, Washington, D.C., 179, 1984.

Beck B.F., A Common Pitfall in the Design of RCRA Ground-Water Monitoring Programs, Ground Water, Volume 21, No. 4, p. 488-489, 1983.

Barcelona, M.J., J.P. Gibb, J.A. Helfrich and E.E. Garske, Practical Guide for Ground-Water Sampling; Illinois State Water Survey, SWS Contract Report 374, Champaign Illinois, 93, 1985.

Brobst, R.D. and P.M. Buszka, The Effect of Three Drilling Fluids on Ground-Water Sample Chemistry, Groundwater Monitoring Review, Volume 6, No. 1, 62-70, 1986.

Bryden, G.W., W.R. Mabey and K.M. Robine, Sampling for Toxic Contaminants in Ground Water, Ground-Water Monitoring Review, Vol. 6, No. 2, 67-72, 1986.

Cherry, J.A, R.W. Gillham and J.F. Barker, Contaminants in Ground Water: Chemical Processes, Ground-Water Contamination, Studies in Geophysics, National Academy Press, Washington, D.C., 179, 1984.

Environmental Protection Agency, Final Comprehensive State Ground Water Protection Program Guidance, PB93-163087, 1992.

Environmental Protection Agency, Guidance for Planning for Data Collection in Support of Environmental Decision Making Using the Data Quality Objective Process, EPA QA/G-4, 1993.

Heath, R.C., Ground-Water Regions of the United States, United States Geological Survey Water Supply Paper 2242, Superintendent of Documents, United States Government Printing Office, Washington, D.C., 78, 1984.

Hinchee, R.E. and H.J. Reisinger, Multi-Phase Transport of Petroleum Hydrocarbons in the Subsurface Environment: Theory and Practical Application, Proceedings of the NWWA/APA Conference on Petroleum Hydrocarbons and Organic Chemicals in Ground Water: Prevention, Detection and Restoration, National Water Well Association, Dublin, Ohio, 58-76, 1985.

Huber, W.F, The Use of Downhole Television in Monitoring Applications, Proceedings of the Second National Symposium on Aquifer Restoration and Ground-Water Monitoring, National Water Well Association, Dublin, Ohio, 285-286, 1982.

Kovski, J.R., Physical Transport Process for Hydrocarbons in the Subsurface, Proceedings of the Second International Conference on Ground Water Quality Research, Oklahoma State University Printing Services, Stillwater, Oklahoma, 127-128, 1984.

Morahan, T. and R.C. Doorier, The Application of Television Borehole Logging to Ground-Water Monitoring Programs, Ground-Water Monitoring Review, Vol. 4, No. 4, 172-175, 1984.

Nielsen, D.M. and G.L. Yeates, A Comparison of Sampling Mechanisms Available for Small-Diameter Ground-Water Monitoring Wells, Proceedings of the Fifth National Symposium and Exposition on Aquifer Restoration and Ground-Water Monitoring, National Water Well Association, Dublin, Ohio, 237-270, 1985.

Norman, W.R., An Effective and Inexpensive Gas-Drive Ground-Water Sampling Device, Ground-Water Monitoring Review, Vol. 6, No. 2, 56–60, 1986.

Pettyjohn, W.A., Cause and Effect of Cyclic Changes in Ground-Water Quality, Ground-Water Monitoring Review, Vol. 2, No. 1, 43–49, 1982.

Schwarzenbach, R.P. and W. Giger, Behavior and Fate of Halogenated Hydrocarbons in Ground Water, John Wiley and Sons, New York, 446–471, 1985.

Villaume, J.F., Investigations at Sites Contaminated with Dense, Non-Aqueous Phase Liquids (DNAPLs), Ground-Water Monitoring Review, Vol. 5, No. 2, 60–74, 1985.

CHAPTER 4

Equipment Decontamination

When performing environmental investigations, all sampling equipment must be treated as if it is contaminated, and therefore should be thoroughly decontaminated between sampling points. Decontamination is defined as the process of neutralizing, washing, rinsing, and removing exposed outer surfaces of equipment and personal protective clothing to minimize the potential for contaminant migration (EPA 1992), and assures the collection of representative environmental samples. The only way to eliminate decontamination is by using disposable or dedicated sampling equipment.

It is critical to test the effectiveness of any decontamination procedure so that the credibility of environmental samples cannot be questioned. This is accomplished through the preparation of equipment rinsate blank samples which are prepared by pouring distilled/deionized water over decontaminated sampling equipment, and collecting the rinsate in sample bottles. The rinsate sample is then shipped to the laboratory and is analyzed for the same parameters as the environmental sample. If the results identify contamination in the rinsate blank, the decontamination procedure was proved to be ineffective. In this case, the analytical results for the sample collected with the contaminated equipment may need to be rejected.

Each morning prior to sampling, one rinsate blank should be collected from each type of sampling equipment to be used that day. For example, if soil and sediment samples are to be collected, one rinsate sample should be collected from the equipment to be used for soil sampling, and one from the equipment to be used for sediment sampling. If soil samples are to be collected using a hand auger, bowl, and spoon, the rinsate sample should be collected by first placing the spoon in the bowl, then pouring ASTM Type II reagent-grade (or equivalent) water over the hand auger while catching the drippings in the bowl. The water from the bowl is then poured into appropriate sample bottles.

The decontamination of drilling and sampling equipment should be performed as close to the sampling site as possible in order to prevent the possible spread of contamination. Decontamination can be performed by physically removing contaminants, inactivating contaminants by chemical detoxification or disinfection/sterilization, or removing contaminants by a combination of both physical and chemical methods (EPA 1992).

The proper decontamination procedure to be used for a piece of equipment is dependent on the degree to which the equipment may come in contact with the sample to be analyzed. In general, equipment that is used directly for sample collection undergoes the most rigorous decontamination procedure. For example, drilling equipment such as augers, drill bits, and drill rod do not need to be decontaminated to the same degree as sample collection tools such as a split-spoon sampler or Teflon bailer. The following procedures are consistent with procedures recommended by EPA Region VIII, and are very similar to procedures recommended by other EPA guidance documents.

LARGE EQUIPMENT

The following method should be considered for decontamination of large equipment used to assist in the collection of environmental samples. This equipment includes drilling augers, drill bits, drill rod, soil-gas rod, Direct Push rod, Cone Penetrometer equipment, etc. Well casing and screen should also be decontaminated using this method.

Decontamination Procedure

1. Remove soil adhering to augers, drill rod, and other equipment by scraping, brushing, or wiping.
2. Thoroughly pressure wash equipment with potable water and a non-phosphatic laboratory grade detergent using a steam cleaner.
3. Thoroughly rinse equipment with potable water using a steam cleaner.
4. Wrap the equipment in plastic sheeting or aluminum foil to keep it clean prior to use.

The waste material generated during Step 1 should be containerized in drums at the drill site. Steps 2, 3, and 4 should be performed at a decontamination pad where waste water can be collected and containerized. If a permanent decontamination pad is not available near the site (Figure 4.1), inflatable pads can be used (Figure 4.2).

SAMPLING EQUIPMENT

The following methods should be considered to decontaminate all types of equipment used to collect environmental samples. This equipment includes split-spoon and thin-walled tube samplers; knives, spoons, spatulas, trowels, and other hand sampling equipment used to handle soil and sediment samples; and Teflon or stainless steel water sampling tools such as dippers, bailers, downhole pumps, intake and discharge lines, and barrel filters used to collect water samples.

When samples are to be analyzed for inorganics, including radionuclides, a dilute hydrochloric or nitric acid solution rinse must be used in the decontami-

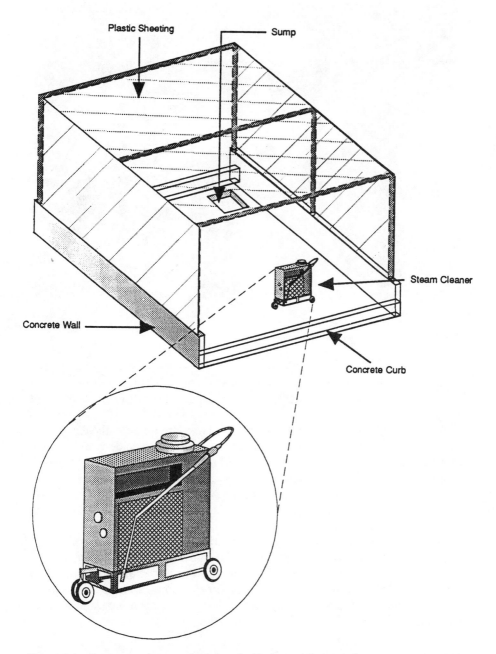

Figure 4.1.   Permanent decontamination pad with sump and steam cleaner.

Figure 4.2. Temporary decontamination pad.

nation procedure. Dilute hydrochloric acid is generally preferred over nitric acid when cleaning stainless-steel because nitric acid may oxidize this material (EPA 1992).

When samples are to be analyzed for organic contaminants, the equipment decontamination procedure should include a rinse with pesticide-grade acetone and hexane, in that order (EPA 1992). Many large sampling programs use pesticide-grade methanol instead of acetone.

When samples are to be analyzed for both inorganics and organics, a dilute hydrochloric or nitric acid solution rinse must be used in combination with a pesticide-grade acetone and hexane rinse.

Below are three equipment decontamination procedures. The first procedure should be used when samples are to be analyzed for only inorganic and/or radiological parameters only. The second procedure should be used when samples are to be analyzed for organic parameters only, and the third should be used if samples are to be analyzed for both organic and inorganic and/or radiological parameters.

Inorganic/Radiological Procedure

1. Remove soil adhering to sampling equipment by scraping, brushing, or wiping.

2. Wash thoroughly with a strong nonphosphate detergent/soap wash water.
3. Rinse thoroughly with tap water.
4. Rinse with ASTM Type II (or equivalent) water.
5. Rinse with dilute hydrochloric or nitric acid solution.
6. Rinse with ASTM Type II (or equivalent) water.
7. Place equipment on a clean piece of aluminum foil and allow to air dry.

Organic Procedure

1. Remove soil adhering to sampling equipment by scraping, brushing, or wiping.
2. Wash thoroughly with a strong nonphosphate detergent/soap wash water.
3. Rinse thoroughly with tap water.
4. Rinse with ASTM Type II (or equivalent) water.
5. Rinse with pesticide-grade acetone (or methanol).
6. Rinse with pesticide-grade hexane.
7. Place equipment on a clean piece of aluminum foil and allow to air dry.

Combined Inorganic/Radionuclide/Organic Procedure

1. Remove soil adhering to sampling equipment by scraping, brushing, or wiping.
2. Wash thoroughly with a strong nonphosphate detergent/soap wash water.
3. Rinse thoroughly with tap water.
4. Rinse with ASTM Type II (or equivalent) water.
5. Rinse with dilute hydrochloric or nitric acid solution.
6. Rinse with ASTM Type II (or equivalent) water.
7. Rinse with pesticide-grade acetone (or methanol).
8. Rinse with pesticide-grade hexane.
9. Place equipment on a clean piece of aluminum foil and allow to air dry.

Once a decontaminated piece of sampling equipment has been allowed to air dry, it should be wrapped tightly in aluminum foil (shiny side facing out) to keep it clean prior to use. Prior to wrapping equipment that was used in a radiological environment, it should be screened using an appropriate instrument to assure that the decontamination procedure was effective. Rinsate blanks are also used as a check on the effectiveness of the decontamination procedure (see page 210).

The dilute acid solution, acetone, hexane, and distilled/deionized water are most easily handled when they are contained within Teflon squirt bottles. Squirt bottles made of substances other than Teflon should not be used, since they have the potential to contaminate the solutions. When using these solutions, the drippings should be caught in a bucket or tub. The acetone and hexane drippings should be allowed to volatilize into the air, while the acid solution drippings should be neutralized using baking soda.

A decontamination line should be set up as shown in Figure 4.3. Tub 1 and 2 are to be used for the soap wash and clean water rinse, respectively. Tub 3 is used to collect acid and solvent rinse solutions. The table is used to set the decontaminated equipment on, to allow it to air dry. The decontamination line should never be set up downwind of any sampling operations, since contaminants carried in the air could contaminate the equipment. The decontamination line is preferably set up crosswind of the sampling operations.

Decontaminating pumps is more difficult than other sampling equipment since many of them are not easy to disassemble. The procedure to use for these tools is to place the pump in a large decontamination tube full of nonphosphate detergent and tap water. The pump is turned on, which forces the wash water through the pump and discharge line. The same procedure is repeated in a tub full of clean tap water. The outside of the pump and outside of the pressure and discharge lines are then decontaminated using the same methods described above.

All of the water generated from this decontamination procedure must be

Figure 4.3. Decontamination line.

containerized and then analytically tested to determine the appropriate disposal procedure. Discharge permits must be acquired from the city and/or county prior to discharging decontamination water into any sanitary or storm sewer system.

REFERENCES

Environmental Protection Agency, A Compendium of Superfund Field Operations Methods, EPA/540/P-87/001a, 1987.

Environmental Protection Agency Region VIII, Draft Standard Operating Procedures For Field Samplers, Rev. 4, 1992.

Environmental Protection Agency, RCRA Facility Investigation (RFI) Guidance, PB89-200299, 1989.

CHAPTER 5

Sample Preparation, Documentation, and Shipment

After the sample collection procedure is complete, sample containers must be preserved, capped, custody sealed, and shipped along with appropriate documentation to the analytical laboratory. Great care should be taken when preparing samples for shipment since an error in this procedure has the potential of invalidating the samples and subsequent data.

SAMPLE PREPARATION

Immediately after a sample bottle has been filled, it must be preserved as specified by QAPP or specific analytical laboratory requirements. Sample preservation requirements vary, based on the sample matrix and the analyses being performed. The only preservation typically required for soil or sediment samples is generally temperature cooling to 4°C. For water samples, some analyses only require cooling to 4°C, while others also require a chemical preservative such as nitric acid (HNO_3), sulfuric acid (H_2SO_4), hydrochloric acid (HCl), or sodium hydroxide (NaOH). Enough acid or base is added to the sample bottle to either lower the pH to below 2, or raise the pH above 10. The chemicals used to preserve a sample must be of analytical grade to avoid the potential for contaminating the sample.

To avoid any difficulties associated with adding chemical preservatives to sample containers in the field, it is recommended that these preservatives be added to sample bottles in a controlled setting prior to entering the field. This alternative reduces the chances of improperly preserving sample bottles or introducing field contaminants into a sample bottle while adding the preservative.

The preservative should be transferred from the chemical bottle to the sample container using either a disposable polyethylene pipet or a standard glass pipet. A glass eye dropper with rubber bulb is not recommended since the rubber has a potential of introducing contaminants into the sample.

The disposable pipet is made of polyethylene, and should be used only once and then discarded. This pipet is more convenient than the standard glass pipet method and provides the least opportunity for the cross-contamination of samples. The standard glass pipet is preferred over the disposable pipet when bottles for volatile organic analysis need to be preserved, since polyethylene has the potential of providing trace volatile organics to the sample.

The volume of the preservative needed to lower or raise the pH to the appropriate level should be determined before the first sample is collected. The pH of the first preserved sample bottles should be checked by pouring a few drops of water from each bottle onto a strip of litmus paper. The litmus paper should never be dipped into the bottle since this may contaminate the sample. Spot checks should be performed regularly to confirm that samples are being adequately preserved.

After a sample container has been preserved, a Teflon-lined cap or lid is screwed on tightly to prevent the container from leaking. The sample label is filled out, noting the sampling time and date, sample identification number, sampling depth, analyses to be performed, sampler's initials, etc. A custody seal is then placed over the cap or lid just prior to placing the sample bottle into the sample cooler. The custody seal is used to detect any tampering with the sample prior to analysis.

DOCUMENTATION

Accurate documentation is essential for the success of a field sampling program. It is only through documentation that a sample can be tied into a particular sampling time, date, location, and depth. Consequently, field logbooks must be kept by every member of the field team, and should be used to record information ranging from weather conditions, to the time the driller stubbed his right toe. To assist the documentation effort, standardized forms are commonly used to outline the information which needs to be collected. Some of the more commonly used forms include:

- borehole log forms
- well completion forms
- well development forms
- well purging/sampling forms
- water level measurement forms
- soil-gas measurement forms
- instrument calibration forms, and
- sample log form.

Other documentation needs associated with sample identification and shipment include:

- sample labels
- chain-of-custody form
- custody seals, and
- shipping airbill.

In addition to the above documentation requirements, a careful file must be kept to track important information such as:

- field variances
- equipment shipping invoices
- sample bottle lot numbers
- documented purity specifications for preservatives, distilled water, and calibration liquids and gases
- instrument serial numbers, and
- QA nonconformance notices.

Field Logbooks

Field logbooks are intended to provide sufficient data and observations to enable participants to reconstruct events that occurred during projects and to refresh the memory of the field personnel if called upon to give testimony during legal proceedings. In a legal proceeding, logbooks are admissible as evidence, and consequently must be factual, detailed, and objective (EPA 1987).

Field logbooks must be permanently bound, the pages must be numbered, and all entries must be written with permanent ink, signed, and dated. If an error is made in the notebook, corrections can be made by the person who made the entry. A correction is made by crossing out the error with a single line, so as not to obliterate the original entry, and then entering the correct information. All corrections must be initialed and dated (EPA 1985, 1987, 1992).

Observations or measurements that are taken in an area where the logbook may be contaminated can be recorded in a separate bound and numbered logbook before being transferred into the master field notebook. All logbooks must be kept on file as permanent records, even if they are illegible or contain inaccuracies that require a replacement document (EPA 1987).

The first page of the logbook should be used as a Table of Contents to facilitate the location of pertinent data. As the logbook is being completed, the page number where important events can be found should be recorded. The very next page should begin recording daily events. The first daily event entry should always be the date, followed by a detailed description of the weather conditions. All of the following entries should begin with time that the entry was made. Any space remaining on the last line of the entry should be lined out to prevent additional information being added in the future. At the end of the day, any unused space between the last entry and the bottom of the page should be lined out, signed, and dated, to prevent additional entries being made at a later date.

To assure that a comprehensive record of all important events is recorded, each team member should keep a daily log. The Field Manager's logbook should record information at the project level, and should record information such as:

- time when team members, subcontractors, and the client arrive or leave the site
- names and company affiliation of all people who visit the site
- summary of all discussions and agreements made with team members, subcontractors, and the client
- summary of all telephone conversations
- detailed explanations of any deviations from the Field Sampling Plan, noting who gave the authorization, and what paperwork was completed to document the change
- detailed description of any mechanical problems which occurred at the site, noting when and how it occurred, and how it is being addressed
- detailed description of any accidents that occur, noting who received the injury, how it occurred, how serious the injury was, how the person was treated, and who was notified
- other general information such when and how equipment was decontaminated, what boreholes were drilled, and what samples were collected that day.

The team members' logbooks should record information more at the task level. Examples of the types of information that should be recorded in these logbooks are:

- the time when drilling began and ended on a particular borehole
- the level of personal protective equipment used at the site
- sample collection times for all samples collected
- the total depth of the borehole
- detailed description of materials used to build monitoring wells, including: type of casing material used; screen slot size; length of screen; screened interval; brand name, lot number, and size of sand used for the sand pack; brand name, lot number, and size of bentonite pellets used for the bentonite seal; brand name and lot number of bentonite powder and cement used for grout; well identification number
- details on when, how, and where equipment was decontaminated, and what was done with the wastewater
- description of any mechanical problems that occurred at the site, noting when and how it occurred and how it was addressed
- summary of all discussions and agreements made with other team members, subcontractors, and the client
- summary of all telephone conversations, and

- detailed description of any accidents that occur, noting who received the injury, how it occurred, how serious the injury was, how the person was treated, and who was notified.

It is essential that each member of the field team record as much information as possible in their logbooks, since this is generating a written record of the project. Years after the project is over, these notebooks will be the only means of reconstructing events that occurred. With each team member recording information, it is not uncommon for one member to record information that another member missed.

For all photographs taken at a site, a photographic log should be kept. This log should record the date, time, subject, frame, roll number, and photographer. For "instant photos," the date, time, subject, and photographer should be recorded directly on the developed picture. A clear photograph of a sample jar, showing the label, custody seal, and the color and amount of sample can be very useful in reconciling any later discrepancies (EPA 1987).

Field Sampling Forms

It is recommended that standardized field sampling forms be used to assist the sampler in a number of field activities. The most commonly used forms include the Borehole Log Form, Well Completion Form, Well Development Form, Well Purging/Sampling Form, Water Level Measurement Form, Soil-Gas Measurement Form, and Instrument Calibration Form (Figures 5.1 through 5.6). Forms are commonly used to reduce the amount of documentation required in the logbook. Forms are also effective in reminding the sampler of what information needs to be collected, and makes it more obvious when the necessary information was not collected.

When forms are used, they must be permanently bound in a notebook, the pages must be numbered, and all entries must be written with permanent ink. If an error is made in the notebook, corrections can be made by the person who made the entry. A correction is made by crossing out the error with a single line so as not to obliterate the original entry, and then entering the correct information. All corrections must be initialed and dated (EPA 1985, 1987, 1992). The person who completed the form must sign and date the form at the bottom of the page. It is recommended that the Field Manager also sign the form to confirm that it is complete and accurate.

Identification and Shipping Documentation

The essential documents for sample identification and shipment include the sample label, custody seal, Chain-of-Custody Form, and shipping airbill. Together these documents allow samples to be shipped to an analytical laboratory under custody. If custody seals are broken when the laboratory receives the samples, the assumption must be made that the samples were tampered

Science Applications International Corporation
An Employee-Owned Company

Borehole # _____

page _____ of _____

Borehole Log Form

Project _____	Total Depth _____		START	FINISH
Location _____	Borehole Diameter_____	Date	_____	_____
Geologic Log by _____	Depth to Water_____	Time	_____	_____
Driller _____	Rig _____	How Left _____		
Geophysics by _____	Bit(s)_____			
Weather _____	Drilling Fluid _____			

Depth — 0 —	Pene. Rate/ Blow Cts	Circu- lation Q (gpm)	OVA/ HNU	Sample			Geologic and Hydrologic Description	% Core Recovery
				#	Inter- val	Lith. Symbol		

4-111692-238

Figure 5.1. Example of a Borehole Log Form used to record borehole lithology information.

**Science Applications
International Corporation**
An Employee-Owned Company

Well Completion Form

Location: _____ Elevation: Ground Level _____
Personnel: _____ Top of Casing _____

DRILLING SUMMARY:	CONSTRUCTION TIME LOG:				

DRILLING SUMMARY:

Total Depth _____
Borehole Diameter _____
_____

Driller _____

Rig _____
Bit(s) _____
_____

Drilling Fluid _____

Surface Casing _____

WELL DESIGN:
Basis:
Geologic Log
Casing String(s): C=Casing S=Screen

_____ _ _____ _____
_____ _ _____ _____
_____ _ _____ _____
_____ _ _____ _____
_____ _ _____ _____
_____ _ _____ _____

Casing: C1 _____
         C2 _____
         C3 _____
         C4 _____
Screen: S1 _____
        S2 _____
        S3 _____
        S4 _____
Centralizers _____
_____

Filter Material _____

Cement _____

Other _____
_____
_____

CONSTRUCTION TIME LOG:

Task	Start Date	Start Time	Finish Date	Finish Time
Drilling:				
Geophys. Logging:				
Casing: _____				
Filter Placement:				
Cementing:				
Development:				
Other: _____				

Comments:
_____
_____
_____
_____
_____
_____
_____
_____

Key:

▨	Bentonite	▦	Sand
▨	Cement/Grout	▦	Silt
▨	Sand Pack	▨	Clay
▨	Drill Cuttings	▤	Screen
▨	Gravel		

4-111692-238

Figure 5.2.   Example of a Well Completion Form.

Science Applications
International Corporation
An Employee-Owned Company

Well Development Form

Project Name and Number: _____

Well Number and Location: _____

Development Crew: _____ Driller (if applicable): _____

Water Levels/Time: Initial: _____ Pumping: _____ Final: _____

Total Well depth: Initial: _____ Final: _____

Date and Time: Begin: _____ Completed: _____

Development: Method(s): _____

Total Quantity of Water Removed: _____ gallons

Date/Time and Pump Setting	Discharge Rate* and Measurement Method	Field Measurements				Remarks
		Temp (°C)	Specific Conductivity (umhos/cm)	pH (Standard Units)	Turbidity	

*gallons per minute or bailer capacity   4-111692-238

Figure 5.3. Example of a Well Development Form.

**Science Applications
International Corporation**
An Employee-Owned Company

Well Purging and Sampling Form

Site _____ Well No._____

Date(s) _____ Geologist _____

Purging Bailer _____ Equipment Used _____

Sampling Bailer _____ Measurement Reference Datum _____

DATA FROM IMMEDIATELY BEFORE AND AFTER DEVELOPMENT:

Depth to water measured from TOC (ft): Before Purging: _____ { After Purging: _____

Total purging time (min): _____   { After Sampling: _____

Depth to sediment in well (ft): Before Purging: _____   After Purging: _____

	Time Since Purging Started (min)	Time	Cumulative Volume Removed	Water Temp °C	pH of Water	Conductivity (μ mhos/cm)	Turbidity (NTUs)	Water Appear-ance*	Date
Before									
During									
During									
During									
During									
During									
During									
After									

*CL = clear   CO = cloudy   TU = turbid

Comments
_____
_____
_____
_____
_____
_____
_____
_____
_____
_____
_____

4-111692-238

Figure 5.4.   Example of a Well Purging and Sampling Form.

**Science Applications
International Corporation**
An Employee-Owned Company

Water Level Measurements

Measurement Team: _____

Project Number and Location: _____

Measuring Instrument: _____

Well No.	Date	Time	Tape Reading		Depth to Water (ft)	Initials	Remarks
			Measure Pt.	Water Level			

Measuring Point: Point where measurement was taken. Top of casing (TOC); Top of Protective Steel Casing (TOSC); Land Surface (LS), etc.

Depth to Water: Measurements should be recorded to the nearest 0.01 ft.

Remarks: Any conditions that may influence the water level measurements.

4-111692-238

Figure 5.5. Example of a Water Level Measurement Form.

**Science Applications
International Corporation**
An Employee-Owned Company

Soil-Gas Measurement Form

Project Name and Number: —————————————————————————

Site: ——

Weather Conditions: ——————————————————————————————

Background Reading: ——————————————————————————————

Date	Time	Measuring Device	Reading	Units	Initials	Comments

4-111692-238

Figure 5.6. Example of a Soil-Gas Measurement Form.

with during shipment. Consequently, the samples would need to be collected over again.

Sample Labels

The primary objective of the sample label is to link a sample bottle to a sample number, sampling date and time, and analyses to be performed. The sample label in combination with the Chain-of-Custody Form is used to inform the laboratory what the sample is to be analyzed for. At a minimum, a sample label should contain the following information (Figure 5.7):

- sample number
- sampling time and date
- analyses to be performed
- preservatives used
- sampler's initials
- name of the company collecting the sample, and
- name and address of the laboratory performing the analysis.

Figure 5.7. Example of a sample label and Chain-Of-Custody Seal.

To save time in the field, and to avoid the potential for errors, all of the above information should be added to the sample label before going into the field, with the exception of the sampling time and date, and sampler's initials. This information should be added to the label following the capping of the sample bottle, immediately after sample collection, and should reflect the time that sampling began, as opposed to the time sampling was completed.

An effective sample numbering system is a key component to any field sampling program since it serves to tie the sample to its sampling location. The problem with using a simple numbering system such as 1, 2, 3, . . . is that the number tells you nothing regarding the location, depth, or sample media.

CLP (Contract Laboratory Program) was developed for sites on the National Priorities List (NPL), and allows sample numbers to be no more than six digits in length for metals analyses, and nine digits for organics. If samples are to be analyzed for both metals and organics, the six-digit sample number must be used. Since CLP analyses are not required for non-NPL sites, sample numbers with 10 or more digits can be used at these sites. However, it is important to coordinate the proposed sampling numbering scheme with the laboratory to identify any limitations they may have.

Even with only six digits, an effective numbering system can be developed with a little imagination. As an example, assume that a project has a total of nine sites, where shallow soil, deep soil, sediment, surface water, groundwater, and concrete core samples are to be collected. From each sampling location, no more than nine samples are to be collected.

For the project outlined above, each of the six sample number digits could be used to represent the following:

First digit: Site Number (1,2,3. . .9)

1 = Gas Station
2 = Fire Training Area
3 = Fuel Tank Farm
4 = Fuel Decontamination Facility
5 = Power Plant
6 = Missile Silo 576-D
7 = Missile Silo 395-C
8 = Space Launch Complex 3D
9 = Space Launch Complex 4D

Second digit: Media Number (1,2,3. . .6)

1 = shallow soil
2 = deep soil
3 = sediment
4 = surface water
5 = groundwater
6 = concrete core

Third, fourth, and fifth digit: Sampling Location Number

    BH9 = Borehole (1,2,3. . .9)
    MW9 = Monitoring Well (1,2,3. . .9)
    CH9 = Corehole (1,2,3. . .9)

Sixth digit: Sample number

    1 = first sampling interval
    2 = second sampling interval
    3 = third sampling interval
    4 = fourth sampling interval
    5 = fifth sampling interval
    6 = sixth sampling interval
    7 = seventh sampling interval
    8 = eighth sampling interval
    9 = ninth sampling interval

The sample number "32BH43" would therefore indicate that the sample was collected from the Fuel Tank Farm, it is a deep soil sample, it was collected from Borehole No. 4, and it was collected from the third sampling interval. For a larger sampling program, the borehole, monitoring well, and corehole designations could be shortened to free up one of the digits for another use. When using both numbers and letters in a sample number, one should try to avoid using letters such as S, l, and O which can easily be misinterpreted as the numbers 5, 1, and 0.

When non-CLP analyses are being performed, sample numbers as long as 10 or 12 characters are commonly used. These additional characters should be taken advantage of, since the easier it is to interpret the sample identification number, the easier it will be to interpret the data as reported by the laboratory. An example of an effective nonCLP sample numbering system is as follows:

    Sample Number: V-FTF-B4-DS-03
        V = Vandenberg Air Force Base
        FTF = Fuel Tank Farm
        B4 = Borehole No. 4
        DS = Deep Soil
        03 = Third sampling interval.

Chain-of-Custody Forms and Seals

Chain-of-custody is the procedure used to document who is responsible for the sample from the time it is collected to the time it is analyzed at the laboratory. After a sample bottle has been filled, preserved, and labeled, a custody seal is signed and dated, then placed over the bottle cap to assure that the sample is not tampered with (Figure 5.7). The custody seal is a fragile piece of tape designed to break if the bottle cap is turned or tampered with.

The person who signs their name to the custody seal automatically assumes ownership of the sample until the time it is packaged and custody is either signed over to the laboratory or to an overnight mail service. This transfer of custody is recorded on a Chain-of-Custody (COC) Form (Figure 5.8). The COC Form is also the document used by the laboratory to identify which analyses to perform on which samples. When the laboratory receives a sample shipment, they assume responsibility, or custody of the samples by signing the COC Form. The sample bottles in the shipment are then counted, and the requested analyses on the sample bottles are compared against the requested analyses on the COC Form. If there is an inconsistency, the laboratory should contact the Project Manager or Field Manager for clarification. These communications should all be carefully documented.

An overnight shipping airbill can be used to document the transfer of custody from the field to the mail service. If this option is selected, the shipper's airbill number must be recorded on the COC Form, which can be taped inside the sample container. When the laboratory signs the shipper's delivery form and the COC Form inside the sample container, they have assumed custody of the sample.

Other Important Documentation

Careful documentation should also be maintained for all incoming shipments of materials and supplies. The most critical documents to keep on file include: shipping invoices; sample bottle lot numbers; purity specification for preservatives, distilled water, and calibration liquids and gases; and instrument serial numbers and calibration logs.

Copies of shipping invoices can be used by the Project Manager to keep track of equipment costs, and can expedite the return of malfunctioning equipment. It is critical to keep track of sample bottle lot numbers and purity specifications for all chemicals used, since improperly decontaminated bottles and low quality preservation or decontamination chemicals can contaminate samples. Instrument serial numbers are recorded to assist in tracking which instruments are working well, and which must undergo repair.

SAIC Science Applications International Corporation
An Employee-Owned Company

Chain of Custody Record

Laboratory Log Number ________

Client Name	Job Number or Purchase Order Number
Project Name	
Project Manager	Sampler(s)

Sample Number	Date Sampled	Time Sampled	Matrix Type	Sample Description	Number of containers	Analyses required							Hazardous sample Special handling required	Remarks

Signature	Company	Date	Time
Relinquished by			
Received by			
Relinquished by			
Received by			
Relinquished by			
Received by			4-111692-238

Figure 5.8. Example of a Chain-Of-Custody Form.

REFERENCES

Environmental Protection Agency, Guidance on Remedial Investigations Under CER-CLA, EPA/540/G-85/002, 1985, pp. 4-2, 4-4.

Environmental Protection Agency, A Compendium of Superfund Field Operations Methods, EPA/540/P-87/001a, 1987, pp. 4-13.

Environmental Protection Agency Region VIII, Draft Standard Operating Procedures For Field Samplers, 1992, p. 27.

Environmental Protection Agency, RCRA Facility Investigation (RFI) Guidance, PB89-200299, 1989.

CHAPTER 6

Health and Safety

As a result of the passing of RCRA in 1976 and CERCLA in 1980, numerous jobs were created in the hazardous waste industry. In the process of creating these jobs, another dilemma had been created. Laws were lacking to protect the workers' health and safety, thereby putting this group at the mercy of their employer's safety policies. This void was addressed in 1986 when the President signed SARA into law, which addresses the need to protect employees' exposure to hazardous wastes. Specific safety requirements are outlined under 29 CFR 1910.120.

The development of a Project Health and Safety (H&S) Plan is the single most important element in protecting the health and safety of the field worker. It is only through thorough planning that potential hazards may be anticipated, and steps taken to minimize the risk of harm to the workers. The primary objective of the H&S Plan is to identify, evaluate, control safety and health hazards, and provide for emergency response. The specific elements that the plan should address include:

- a hazard risk analysis
- employee training requirements
- personal protective equipment
- medical surveillance requirements
- frequency and type of air monitoring
- site control plan
- decontamination
- emergency response
- confined-space entry procedure, and
- spill containment procedure.

Prior to writing the H&S Plan, an offsite data-gathering effort should be conducted to identify all suspected conditions that may pose inhalation or skin absorption hazards that are immediately dangerous to life or health (IDLH), or other conditions that may cause death or serious injury. An onsite survey should then be performed using appropriate protective equipment. After the site has been determined to be safe for the initiation of field activities, a monitoring program should be implemented to continue to ensure the worker's safety.

After the H&S Plan has been written, it is essential that all field personnel, including subcontractors, be required to read it. The Field Manager should document that workers have read and understood the H&S Plan by having them sign a training record.

The following sections will discuss the various aspects of training requirements, medical surveillance, hazard overview, personal protective equipment, air monitoring, site control, decontamination, handling drums and other equipment.

TRAINING REQUIREMENTS

According to 29 CFR 1910.120, all employees exposed to hazardous substances, health hazards, or safety hazards must be thoroughly trained in:

- safety, health, and other hazards present on the site
- proper use of personal protective equipment
- work practices which can minimize risks from hazards
- safe use of engineering controls and equipment on the site, and
- medical surveillance requirements.

At the time a job is assigned, an employee must receive a minimum of 40 hours of initial instruction offsite, and a minimum of three days of actual field experience under the direct supervision of a trained, experienced supervisor. Workers who may be exposed to unique hazards must be provided with additional training. For example, employees who are responsible for responding to hazardous emergencies must be trained in what emergencies to anticipate, and how to respond to each situation.

Personnel managers who are responsible for supervising employees engaged in hazardous waste operations must meet all the training requirements outlined above, and have at least eight additional hours of specialized training on managing such operations. This training must be received at the time the job is assigned.

All employees and managers are required to receive eight hours of refresher training once a year. This training should provide a refresher summary of the most important topics discussed during the 40-hour training, as well as some new material.

MEDICAL SURVEILLANCE

Field personnel who work in areas where there is potential of exposure to hazardous substances at or above regulatory limits should have a medical examination prior to entering the field, and at least once every 12 months after this time. An additional medical examination should be performed if a worker begins to show signs or symptoms of overexposure, or is leaving the company.

The objectives of the initial medical examination are to:

- determine an individual's fitness to perform the work
- determine the worker's ability to wear protective equipment, and
- provide baseline data for comparison with future medical data.

The fitness of a worker should be determined by a physician by assessing the medical history and by performing a physical exam. The physical exam should be conducted by a licensed physician according to 29 CFR 1910.120. Although there are no specific requirements as to the content of the examination, it is recommended that it involve an assessment of at least the following:

- height, weight, temperature, pulse, respiration, and blood pressure
- head, nose, and throat
- eyes, including vision tests that measure refraction, depth perception, and color vision
- ears, including audiometric tests. Ear drum should also be checked for perforation
- chest, including heart and lungs
- peripheral vascular system
- abdomen and rectum
- spine and other components of the musculoskeletal system
- genitourinary system
- skin, and
- nervous system.

Tests which should be performed in conjunction with the physical exam include:

- blood
- urine, and
- chest X-ray.

The type of protective equipment that the worker can wear without endangering his/her health must be determined by the physician. Workers who have severe lung disease, heart disease, or back problems should never be put in a position where they need to wear protective equipment. According to 29 CFR 1910.134, a special written assessment of a worker's capacity to perform work while wearing a respirator must be performed before beginning work.

Medical screening is used not only to determine one's fitness level or ability to wear protective equipment, but is also used to establish baseline data. Baseline data is information which is later used to evaluate if any exposures have occurred in the field.

After the baseline examination, annual medical monitoring should be performed. The content of the annual examination is determined by the physician and may vary from partial to full medical examination. In the case of potential overexposure, a full examination should always be performed.

Termination examinations are typically performed when an employee leaves a company. The purpose of the exam is to document the health of the individual at that point in time. If the individual should become contaminated several years later while performing work for another company, the termination examination records would prevent the individual from suing the previous company.

HAZARD OVERVIEW

Hazardous waste sites pose a multitude of health and safety concerns, any one of which could result in serious injury or death. Some of the hazards of greatest concern include:

- electrical hazards
- chemical exposure
- radiological exposure
- oxygen deficiency
- fire and explosion
- heat stress
- noise, and
- heavy equipment accidents.

Hazardous waste sites are particularly dangerous because the identity of substances present are often unknown, the typical disorderly nature of the sites makes them unpredictable, and the need to wear bulky protective clothing commonly adds to the hazard.

Overhead power lines, downed electrical wires, and buried cables all pose a danger of shock or electrocution. To minimize the hazard posed by working in this environment it is important to:

- wear special clothing designed to protect against electrical hazard as described by 29 CFR 1910.137
- use low-voltage equipment with ground-fault interrupters and watertight, corrosion-resistant connecting cables
- monitor weather conditions, and suspend work during electrical storms, and
- properly ground capacitors that may retain a charge before handling.

When working in chemical and radiological environments, exposure can be controlled by wearing appropriate protective clothing, using health and safety monitoring equipment, and by minimizing the length of exposure. Before beginning work in these environments, a health and safety plan must be written that addresses all of the safety concerns, including how to respond in case of an emergency. All employees should be trained in these procedures prior to beginning work.

Oxygen deficiency hazards can be overcome by avoiding work in confined spaces whenever possible, and by requiring the use of oxygen meters. Fire and explosion hazards can be minimized by controlling the use of open flames, using nonsparking tools, and using an explosivity meter in potentially explosive environments. Fire escape routes should be identified before beginning work.

When work is being performed in hot and humid environments, heat stress is a common hazard. To avoid this condition, the Health and Safety Officer is responsible for seeing that workers take regular breaks, recharge their body fluids, and wear appropriate protective equipment.

Whenever noise is a potential hazard, employees should be required to wear appropriate hearing protection. When working around heavy equipment, accidents can be avoided by wearing personal protective equipment such as hard hats, safety glasses, and steel-toed boots; using two-way radios to improve communication; and by working at a steady, unrushed pace.

AIR MONITORING

The health and safety of the site worker may be jeopardized by airborne contaminants. This danger can be controlled by using screening instruments to monitor the air.

Most air screening instruments used for H&S are designed to measure gases and vapors in the air, and provide immediate monitoring results. These instruments are used by field samplers to provide early warning signs of a hazardous environment. Many of them have the capability of measuring in the ppm range. Some of the advantages of these direct-reading instruments include the ability to provide immediate information regarding explosive, flammable, and oxygen-deficient environments, and provide information on a specific gas or a general screen of gases.

Some of the most common direct-reading instruments include the Combustible Gas Indicator (CGI), Flame Ionization Detector (FID), Photoionization Detector (PID), colorimetric indicator tube, and oxygen indicators.

The CGI measures the concentration of a flammable vapor or gas in air and presents the results as a percentage of the lower explosive limit (LEL) of the calibration gas. Most CGI meters operate on the "hot wire" principle. In the combustion chamber is a platinum wire that is heated. The filament is an integral part of a balanced resistor circuit called a Wheatstone Bridge. The hot filament combusts the gas on the immediate surface of the element, thus raising the temperature of the filament. As the temperature of the filament increases, so does its resistance. This change in resistance causes an imbalance in the Wheatstone Bridge. This is measured as the ratio of combustible vapor present compared to the total required to reach the LEL. If a gas concentration greater than the LEL and lower than the Upper Explosive Limit (UEL) is present, then the meter needle will stay beyond the 100% level on the meter, indicating that the ambient atmosphere is readily combustible. Some of the limitations of this instrument include:

- The instrument accuracy is dependent, in part, on the difference between the calibration and sampling temperatures.

- There is the potential for the filament being damaged by certain compounds such as silicones, halides, tetraethyl lead, and oxygen-enriched atmospheres.
- Valid readings are not provided in oxygen-deficient conditions.

The FID detects many organic gases and vapors using ionization as the detection method. Inside the detection chamber, the sample is exposed to a hydrogen flame which ionizes the organic vapors. The positively charged carbon-containing ions which are produced are collected by a negatively charged collecting electrode in the chamber. As the positive ions are collected, an electric current proportional to the hydrocarbon concentration is generated. This current is amplified to cause a needle deflection on the instrument meter. The drawbacks to this instrument include its:

- inability to detect inorganic gases and vapors and certain synthetics
- sensitivity is dependent on the contaminant type
- inability to absolutely identify compounds, and
- sensitivity to cold temperatures.

The PID detects many organic gases and vapors and also has the capability of monitoring some inorganics. This instrument operates on the principle that when ultraviolet radiation strikes a chemical compound, the compound will ionize if the radiation is equal to or greater than the ionization potential of the compound. The PID utilizes an electric pump to pull the gas sample past an ultraviolet source. Constituents of the sample are ionized, which produces an instrument response if the ionization potential is equal to or less than the ionizing energy supplied by the instrument ultraviolet lamp. The drawbacks to this instrument include its:

- inability to detect methane
- inconsistent responses when gases are mixed, and
- sensitivity to high humidity.

The colorimetric indicator tube consists of a glass tube filled with an indicating chemical. The tube is connected to a piston cylinder, or bellows-type pump that pulls a known volume of contaminated air at a predetermined rate through the tube. The contaminant reacts with the indicator chemical in the tube, producing a stain whose length is proportional to the contaminant concentration. Some factors to be aware of when considering the use of colorimetric tubes include:

- Altitude or humidity may affect the accuracy of the tube
- Tubes have a shelf life of approximately 2 years, and
- The pump must be checked for leaks before and after use.

An oxygen indicator is generally composed of an oxygen sensing device, which contains an electrolytic solution, Teflon membrane, and electrodes, and a meter readout. Depending on the type of unit selected, air is either drawn into the sensing device by a pump or aspirator bulb or is allowed to equilibrate with the sensor. Once in contact with the sensor, oxygen molecules diffuse through a Teflon membrane into the electrolytic solution. Reactions between the oxygen and the electrodes produce an electric current which is directly proportional to the sensor's oxygen content. The limitations of this meter include:

- Certain gases, such as oxidants, can affect the readings, and
- Carbon dioxide poisons the detector cell. If the ambient air is more than 0.5% carbon dioxide, the oxygen detector cell must be replaced or rejuvenated.

When working in radiological environments, the inhalation of radioactive dust particles can be avoided by wearing air-filtering respirators. Since there are no radiological instruments which can provide immediate air quality information, respirators should always be worn when working in these environments.

RADIOLOGICAL SCREENING INSTRUMENTS

Radiological screening instruments should always be used to monitor field personnel prior to leaving a radiological environment. The purpose of the screening is to determine if a worker has been contaminated, and to prevent the spread of contamination from the site.

To perform the screen, the probe of the alpha and/or beta meter is passed very slowly over the worker's body, focusing on those areas that have the highest probability of being contaminated. These areas include the soles of the worker's shoes, hands, arms, shoulders, and knees.

If elevated activity levels are identified on a piece of clothing, it must be removed and properly disposed of. If contamination is detected on a person's skin, the contaminated area should be carefully washed with soap and water, and then rescreened to determine if the contamination has been removed. To prevent the skin pores from contracting or expanding, the temperature of the wash water should be approximately equal to body temperature. If the contamination was not removed, the procedure should be repeated. A soft brush can be used, but care should be taken not to imbed the contamination deeper into the skin.

SITE CONTROL

Site control is a requirement of 29 CFR 1910.120, since it is needed to prevent the spread of contamination and to protect outsiders from the site hazards. Site control can most easily be obtained by breaking the site into an Exclusion Zone, Contamination Reduction Zone, and Support Zone (Figure 6.1). A Hot Line and Contamination Control Line should be used to control

SUPPORT ZONE

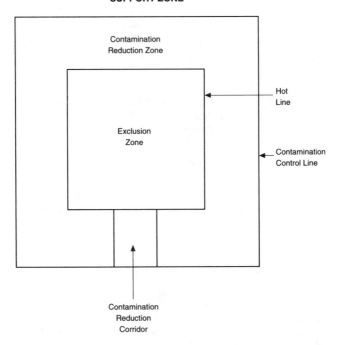

Figure 6.1.   Zones used to control the spread of contamination.

the spread of contamination beyond the boundary of the Exclusion Zone and Contamination Reduction Zone, respectively. These lines should be marked very clearly using signs, placards, or hazard tape.

The Exclusion Zone is the area where the contamination is located at the beginning of the field program, and every effort should be taken to keep the contamination from spreading from this area. This zone can be subdivided into different subareas based on the types of contaminants, or degree of hazard. Subdividing the zone allows more flexibility in safety requirements, operations, and decontamination procedures.

The Contamination Reduction Zone is the area used to decontaminate workers exiting the Exclusion Zone. The contamination level should decrease in this zone as one gets closer to the Support Zone. To reduce the spread of contamination within this zone, a Contamination Reduction Corridor should be used. This corridor should contain all of the contamination reduction wash tubs, and should be the only place where workers can enter and exit the Exclusion Zone. This corridor should have at least two lines of decontamination tubs, one for personnel, and the second for heavy equipment. This area should also be used to store emergency response equipment, fire extinguisher, personal protective equipment, sampling equipment, sampling packaging materials, and a worker temporary rest area.

The Support Zone is the uncontaminated area where workers should not be exposed to any hazardous conditions. Any functions that cannot be performed in a hazardous area are performed here. Personnel in this zone are responsible for alerting the proper agencies in the event of an emergency. All emergency telephone numbers, evacuation route maps, and vehicle keys should be kept in this zone.

PERSONAL PROTECTIVE EQUIPMENT

The purpose of personal protective equipment (PPE) is to shield or isolate individuals from chemical, physical, or biological hazards that may be encountered while working in a contaminated environment. Various types of PPE are effective in protecting the respiratory system, skin, eyes, face, hands, feet, head, body, and auditory system.

It is mandatory that hazardous waste operations, including site characterization, comply with the Occupational Safety and Health Administration (OSHA) outlined in 29 CFR 1910.120 and 29 CFR 1910.132 through 1910.137. Under 29 CFR 1910.120, OSHA requires that all employees working in potentially hazardous environments be trained in how to safely use PPE prior to beginning their jobs. This requirement has resulted in a dramatic reduction in the number of injuries that occur each year.

The EPA has established four levels of worker protection. These range from Level A protection which is designed to protect workers in extremely hazardous and potentially oxygen-deficient environments, to Level D which is designed for very low hazard environments. Within each of these levels, there are recommended types of PPE designed to protect the worker. Tables 6.1 through 6.4 present recommended and optional PPE for each level of protection, in addition to a summary of the protection and limitations of each level. The level of protection selected should also take the worker's location and/or job function into consideration. For example, the worker in the Exclusion Zone may require Level A protection, while Level B or C protection may be adequate for support staff in the Contamination Reduction Zone.

When choosing chemical protective clothing, it is necessary to keep in mind that no one type of clothing material can protect against all chemicals. Therefore, the clothing should be selected using historical information about the chemicals used at the facility. Similar information for body suits can be obtained from vendors. Table 6.5 presents the hazard rating for a number of common chemicals.

When using PPE, each worker must assume responsibility for monitoring the effectiveness of the clothing and equipment. Problems with the PPE may be realized through:

- degradation
- perception of odors

Table 6.1. Summary of PPE Requirements and Limitations for Level A Personal Protection

LEVEL A

EQUIPMENT:

Recommended:

- Positive-pressure, full-facepiece SCBA or supplied-air respirator with escape SCBA.
- Fully-encapsulating, chemical-resistant suit.
- Inner chemical-resistant gloves.
- Chemical-resistant safety boots/shoes.
- Two-way radio communications.

Optional

- Cooling unit.
- Coveralls.
- Long cotton underwear.
- Hard hat.
- Disposal gloves and boot covers.

PROTECTION PROVIDED:

- The highest available level of respiratory, skin, and eye protection.

SHOULD BE USED WHEN:

- The chemical substance has been identified and requires the highest level of protection for skin, eyes, and the respiratory system based on either:
  - measured (or potential for) high concentration of atmospheric vapors, gases, or particulates.
  - site operations and work functions involving a high potential for splash, immersion, or exposure to unexpected vapors, gases, or particulate of materials that are harmful to skin or capable of being absorbed through the intact skin.
- Substances with a high degree of hazard to the skin are known or suspected to be present, and skin contact is possible.
- Operations must be conducted in confined, poorly ventilated areas until the absence of condition requiring Level A protection is determined.

LIMITS:

- Fully-encapsulating suit material must be compatible with the substances involved.

Table 6.2. Summary of PPE Requirements and Limitations for Level B Personal Protection

LEVEL B

EQUIPMENT:

Recommended:

- Positive-pressure, full-facepiece SCBA or supplied-air respirator with escape SCBA.
- Chemical-resistant clothing (overalls and long-sleeved jacket, hooded, one- or two-piece chemical splash suit; disposable chemical-resistant one-piece suit).
- Inner chemical-resistant gloves.
- Chemical-resistant safety boots/shoes.
- Hard hat.
- Two-way radio communications.

Optional

- Coveralls.
- Disposal boot covers.
- Face shield.
- Long cotton underwear.

PROTECTION PROVIDED:

- The same level of respiratory protection but less skin protection than Level A.
- Minimum level recommended for initial site entries until the hazards have been further identified.

SHOULD BE USED WHEN:

- The type and atmospheric concentration of substances have been identified and require a high level of respiratory protection, but less skin protection. This involves atmospheres:
  - with IDLH concentration of specific substances that do not represent a severe skin hazard.
  - that do not meet the criteria for use of air-purifying respirators.
- Atmosphere contains less than 19.5 percent oxygen.
- Presence of incompletely identified vapors or gases is indicated by direct-reading organic vapor detection instrument, but vapors and gases are not suspected of containing high levels of chemicals harmful to skin or capable of being absorbed through the intact skin.

LIMITS:

- Use only when the vapor or gases present are not suspected of containing high concentrations of chemicals that are harmful to skin or capable of being absorbed through the intact skin.
- Use only when it is highly unlikely that work being done will generate either high concentrations of vapors, gases, particulates or splashes of material that will affect exposed skin.

Table 6.3. Summary of PPE Requirements and Limitations for Level C Personal Protection

LEVEL C

EQUIPMENT:

Recommended:

- Full-facepiece, air-purifying, canister-equipped respirator.
- Chemical-resistant clothing (overalls and long-sleeved jacket, hooded, one- or two-piece chemical splash suit; disposable chemical-resistant one-piece suit).
- Inner and outer chemical-resistant gloves.
- Chemical-resistant safety boots/shoes.
- Hard hat.
- Two-way radio communications.

Optional

- Coveralls.
- Disposal boot covers.
- Face shield.
- Escape mask.
- Long cotton underwear.

PROTECTION PROVIDED:

- The same level of skin protection as Level B, but a lower level of respiratory protection.

SHOULD BE USED WHEN:

- The atmospheric contaminants, liquid splashes, or other direct contact will not adversely affect any exposed skin.
- The types of air contaminants have been identified, concentrations measured, and a canister is available that can remove the contaminant.
- All criteria for the use of air-purifying respirators are met.

LIMITS:

- Atmospheric concentration of chemicals must not exceed IDLH levels.
- The atmosphere must contain at least 19.5 percent oxygen.

Table 6.4. Summary of PPE Requirements and Limitations for Level D Personal Protection

LEVEL D

EQUIPMENT:

Recommended:

- Coveralls.
- Safety boots/shoes.
- Safety glasses or chemical splash goggles.
- Hard hat.

Optional

- Coveralls.
- Disposal boot covers.
- Face shield.
- Escape mask.
- Long cotton underwear.

PROTECTION PROVIDED:

- No respiratory protection.
- Minimal skin protection.

SHOULD BE USED WHEN:

- The atmosphere contains no known hazard.
- Work functions preclude splashes, immersion, or the potential for unexpected inhalation of or contact with hazardous levels of any chemicals.

LIMITS:

- This level should not be worn in the Exclusion Zone.
- The atmosphere must contain at least 19.5 percent oxygen.

Table 6.5.   General Hazard Rating for a Number of Common Chemicals of Concern.

Risk Code	Hazard Rating
Tx	high toxic
T	toxic
Cx	highly corrosive
C	corrosive
X	harmful
Xi	irritant
V	potentially harmful
—	no risk

Chemicals listed as "Tx" can cause death or very serious injury and are more potentially hazardous than those listed as "X" which can also cause injury.

Those listed as "V" are the least potentially hazardous and can cause injury only in relatively large doses.

Information was not available for chemicals for which no risk code is indicated.

Class	Chemical Names and (Synonyms)	Risk Code
102	Acetic acid	Cx
391	Acetone	V
111	Acetyl chloride	C
121	Acrolein (Acrylaldehyde or 2-Propenal)	Tx
102	Acrylic acid	C T
350	Ammonia gas	T
340	Ammonium acetate	Xi
340	Ammonium carbonate	T
380	Ammonium hydroxide, <30%	X
380	Ammonium hydroxide, 30-70%	C
292	Benzene	T Cancer
432	Benzonitrile (Phenyl cyanide)	Tx
112	Benzoyl chloride (Benzoic acid chloride)	C
222	Benzyl acetate	V
370	Boric acid	X
261	Bromodichloromethane (Dichlorobromomethane)	X
315	2-Bromoethanol	Tx
300	2-Butanone peroxide	T
222	n-Butyl acetate	V
311	n-Butyl alcohol (1-Butanol)	X
141	n-Butylamine (1-Aminobutane)	Cx
261	n-Butyl chloride (1-Chlorobutane)	X
340	Calcium hydroxide	Cx
508	Carbon disulfide (Carbon bisulfide)	T
261	Carbon tetrachloride (Tetrachloromethane)	T
261	Chlordane, <70%	Tx
350	Chlorine	T
263	Chlorobenzene	X
261	Chlorodibromomethane (Dibromochloromethane)	X
315	2-Chloroethanol (Ethylene chlorohydrin)	Tx
261	Chloroform (Trichloromethane)	X

Class	Chemical Names and (Synonyms)	Risk Code
265	Chloronaphthalenes	X
370	Chromic acid	Cx Sensitization
316	Creosote	T Cancer
142	Di-n-butylamine (Dibutylamine)	CX
263	1,2-Dichlorobenzene (o-Dichlorobenzene)	X
263	1,3-Dichlorobenzene (m-Dichlorobenzene)	X
263	1,4-Dichlorobenzene (p-Dichlorobenzene)	X
261	1,1-Dichloroethane (Ethylidene dichloride)	X
261	cis-1,2- Dichloroethylene	X
261	cis,trans-1,2-Dichloroethylene	X
261	1,2-Dichloropropane	X
291	Diesel fuel	X
292	Diethyl benzene	X
141	1,3-Dimethylbutylamine	V
241	Dimethyl ether	T Sensitization
280	1,1-Dimethylhydrazine	T
441	2,4-Dinitrotoluene	T
278	1,3-Dioxane	X
278	1,4-Dioxane	T
292	Divinyl benzene	X
275	1,2-Epoxybutane	X
311	Ethanol	V
222	Ethyl acetate	V
292	Ethyl benzene	X
261	Ethylene dibromide (1,2-Dibromoethane)	T Cancer
261	Ethylene dichloride (1,2-Dichloroethane)	T Cancer
350	Fluorine	Tx Cx
121	Formaldehyde, 30-70%	T Sensitization
102	Formic acid	Cx
261	Freon 113	V
292	Gasoline, unleaded	X
292	Gasoline, 40-55% aromatics	X
291	Hexane	X
280	Hydrazine	T C Cancer
370	Hydrochloric acid, 30-70%	C
370	Hydrofluoric acid, 30-70%	Tx Cx
350	Hydrogen fluoride	Tx Cx
330	Iodine, solid	X
311	Isobutyl alcohol	X
312	Isopropyl alcohol (2-Propanol)	V
291	Kerosene	X
311	Methanol (Methyl alcohol)	T

(Fosberg 1988)

Table 6.5 (Contd). General Hazard Rating for a Number of Common Chemicals of Concern.

Class	Chemical Names and (Synonyms)	Risk Code	Class	Chemical Names and (Synonyms)	Risk Code
222	Methyl acetate	V	292	Toluene (Methyl benzene)	X
261	Methyl bromide (Bromoethane)	T C	263	1,2,4-Trichlorobenzene	X
261	Methylene bromide (Dibromomethane)	X	263	1,1,1,-Trichloroethane	X
261	Methylene chloride (Dichloromethane)	X	263	1,1,2-Trichloroethane	X
391	Methyl ethyl ketone	V	315	2,2,2-Trichloroethanol	T
502	Mustard gas	Tx	261	Trichloroethylene	X
293	Naphthalene	X	261	1,2,3-Trichloropropane	X
271	Nicotine	Tx	315	Trifluoroethanol	X Sensiti-zation
370	Nitric acid, Red fuming	Cx	294	Turpentine	V
370	Nitric acid, 30-70%	Cx	222	Vinyl acetate	T Cancer
441	Nitrobenzene	T	267	Vinyl chloride	X
441	Nitroethane	X	292	Xylenes	
350	Nitrogen dioxide	Tx			
350	Nitrogen tetroxide	Tx			
441	Nitromethane	X			
104	Oxalic acid	X			
462	Parathion	T			
265	PCB (Polychlorinated biphenyls)	T Cancer			
316	Pentachlorophenol	T			
291	Pentane	V			
370	Perchloric acid, 30-70%	Cx			
316	Phenol, >70% (Carbolic acid)	T C			
370	Phosphoric acid, >70%	C			
380	Potassium hydroxide	C			
311	n-Propyl alcohol (1-Propanol)	X			
271	Pyridine (Azine)	X			
340	Sodium carbonate	Xi			
215	Sodium cyanide, >30%	Tx C			
340	Sodium fluoride	T			
340	Sodium dichromate, <30%	X Sensiti-zation			
380	Sodium hydroxide, 30-70%	Cx			
340	Sodium hypochlorite	C			
340	Sodium thiosulfate	V			
370	Sulfuric acid, >70%	Cx			
316	Tannic acid	Xi			
263	1,2,4,5-Tetrachlorobenzene	X			
261	1,1,1,2-Tetrachloroethane	T			
261	1,1,2,2,-Terachloroethylene	T			

(Fosberg 1988)

- skin irritation
- unusual residues on PPE
- discomfort
- resistance to breathing
- interference with vision or communication, and
- personal responses such as rapid pulse, nausea, and chest pain.

Prior to using a respirator, or Self Contained Breathing Apparatus (SCBA), the equipment should always be inspected to be certain that the rubber or elastomer parts are pliable, and there are no cracks, pinholes, or disfigurations in the facepiece, headband, and valves.

A worker must be fit tested prior to using a respirator to assure an effective seal. Some physical conditions which can prevent a secure respirator facepiece seal include:

- facial scars
- hollow temples
- prominent cheekbones, and
- dentures or missing teeth.

Whenever these or other conditions prevent a good face seal, it is required that the individual not be allowed to wear a respirator.

A respirator can be fit tested by several methods. The most effective test involves a person wearing the respirator to enter a test atmosphere containing a test agent in the form of an aerosol, vapor, or gas. An analytical instrument is then used to quantitatively measure the concentration of the gas both inside and outside the respirator. Another method involves exposing the worker to an irritant smoke, or odorous vapor. The worker will not smell the vapor through the respirator if there is a good seal. Finally, a negative and positive pressure test can be performed. To perform the negative pressure test, the worker cups the respirator inlets with his/her hands and gently inhales for about 10 seconds. Any inward rush of air indicates a poor fit. To perform the positive pressure test, the worker gently exhales while covering the exhalation valve of the respirator to ensure that a positive pressure is built up. Failure to build a positive pressure indicates a poor fit.

When reusable chemical protective suits are used, they must be inspected regularly for cracks, punctures, and discolorations on the inside of the suit. One problem with reusable suits is that chemicals that have begun to permeate the clothing may not be removed during decontamination. Therefore, they may continue to diffuse through the material toward the inside surface and become a hazard for the next person who uses the suit. When determining if clothing can be reused, one should evaluate permeation rates, toxicity of contaminants, care taken when clothing was decontaminated, and whether decontamination is degrading the material.

REFERENCES

Department of Health and Human Services, Occupational Safety and Health Guidance Manual for Hazardous Waste Site Activities, Washington, D.C., Chapters 3-12, October 1985.

Forsberg, K., and S.Z. Mansdorf, Quick Selection Guide to Chemical Protective Clothing, Van Nostrand Reinhold, New York, 1988.

Science Applications International Corporation, Hazardous Waste Operations and Emergency Response Program Training Manual, Oak Ridge, Tennessee, Chapters 2-8, 18-23, March 1991.

Science Applications International Corporation, HAZWOPER Compliance Training: 24 Hour SARA/OSHA, Chapters 1-12, 1992.

BIBLIOGRAPHY

American Conference of Governmental Industrial Hygienists (ACGIH), Threshold Limit Values and Biological Exposure Indices for 1988-1989, Cincinnati, OH, 1988.

American Industrial Hygiene Association, Industrial Noise Manual, Akron, OH, 1975.

Eller, P.M., et al., NIOSH Manual of Analytical Methods, NIOSH 84-100, February 1984.

Griffin, R.D., Principles of Hazardous Materials Management, Lewis Publishers, Chelsea, MI, 1988.

Hall, Jr., R.M., et al., RCRA Hazardous Wastes Handbook, 7th Edition, Government Institutes, Rockford, MD, 1987.

Klinsky, J., et al., Manual of Recommended Practice for Combustible Gas Indicators and Portable Direct Reading Hydrocarbon Detectors, 1st edition, American Industrial Hygiene Association, Akron, OH, 1980.

National Institute of Occupational Safety and Health, Working in Confined Spaces, NIOSH 80-106, 1980.

National Safety Counsel, Accident Prevention Manual for Industrial Operations: Engineering and Technology, 9th Edition, Chicago, IL, 1988.

Schwope, A.D., et al., Guidelines for the Selection of Chemical Protective Clothing, 3rd Edition, American Conference of Governmental Industrial Hygienists, Cincinnati, OH, 1987.

Shleien, B., M.S. Terpilak, The Health Physics and Radiological Health Handbook, Nucleon Lectern Associates, Olacy, MD, 1984.

U.S. Department of Energy, Health Physics Manual of Good Practices for Reducing Radiation Exposure to Levels that are As Low As Reasonably Achievable (ALARA), PNL-6577, June 1988.

U.S. Environmental Protection Agency, Standard Operating Safety Guides, Office of Remedial Response, Hazardous Response Support Division, Edison, NJ, 1984.

CHAPTER 7

Management of Investigation-Derived Waste

Site characterization studies commonly generate some form of IDW such as soil, sludge, liquid, PPE, sampling materials, and decontamination or sample preservation fluids. If the site under investigation is contaminated, each of the above waste items has the potential of being contaminated. Consequently, great care should be taken to assure that it is properly disposed of. The selected disposal option must be both protective of human health and the environment, and comply with ARARs to the extent practicable (EPA 1991,1992).

PROTECTIVENESS

In determining if a waste disposal option is protective, the following items need to be considered (EPA 1992):

- concentration of the contaminants
- total volume of contaminated material
- affected media (i.e., surface water, soil)
- site assess
- location of the nearest population
- potential worker exposures, and
- potential environmental impacts.

In light of the site-specific conditions, it is necessary to use best professional judgment to determine whether an option is protective. For example, if contamination is known to be present in the IDW, and the site is unsecured and located near a residential area, it would not be protective to return excavated soil to the location where it was derived. A better option would be to store the waste in containers in a secured location onsite, or send it immediately to an offsite disposal facility (EPA 1992).

One must also consider the potential effects of IDW on the surrounding environmental media. For example, pouring contaminated development water on the ground near a well is not a good practice because the water could mobilize any hazardous constituents present in the soil or introduce contaminants into clean soil. Rather, the wastewater should be containerized, treated if necessary, and then released to a sanitary sewer or Publicly

Owned Treatment Works (POTW) after receiving the necessary approval (EPA 1992).

COMPLIANCE WITH ARARS

According to the NCP (NCP 55 FR 8756), RI/FS actions must comply with ARARs "to the extent practicable, considering the exigencies of the situation." Therefore, it is generally not required to obtain a waiver if an ARAR cannot be attained during these actions. Potential ARARs for IDW at CERCLA sites include regulations under RCRA (including both federal and state underground injection control regulations), CWA, CAA, TSCA, and other state environmental laws (EPA 1991, 1992).

If RCRA hazardous waste is identified at a site, certain sections under RCRA Subtitle C hazardous waste regulations may be ARARs. RCRA may be relevant and appropriate even if the IDW is not a RCRA hazardous waste. A waste is hazardous under RCRA if it is listed as such under 40 CFR 261.31 – 261.33 or if it exhibits one of four characteristics including ignitability, corrosivity, reactivity, or toxicity. It should be noted that there is no need to treat waste at a CERCLA site as a listed or characteristic RCRA hazardous waste unless test results, records, or other knowledge indicate it as such (EPA 1992).

If IDW, in the form of an aqueous liquid, is considered a RCRA hazardous waste, it should be determined whether the Domestic Sewage Exclusion (DSE) applies to the discharge of that IDW to a POTW. The RCRA DSE exempts domestic sewage and any mixture of domestic sewage and other wastes that passes through a sewer system to a POTW for treatment from classification as a solid waste and, therefore, as a RCRA hazardous waste (40 CFR 261.4, EPA 1992).

If the waste is determined to be a RCRA hazardous waste it is subject to land disposal restrictions (LDRs). Land disposal of IDW is prohibited unless specified treatment standards are met (NCP 55 FR 8759, March 8, 1990). Land disposal occurs when (EPA 1992):

- wastes from different Areas of Contamination (AOCs) are consolidated into one AOC
- wastes are moved outside an AOC (for treatment or storage) and returned to the same or a different AOC, or
- when wastes are excavated, placed in a separate hazardous waste management unit such as an incinerator or tank within the AOC, and then redeposited into the AOC.

Recent changes in these regulations however, allow the establishment of corrective action management units (CAMUs), which are areas within a facility designated by the Regional Administrator for the purpose of implementing corrective action requirements under RCRA §3008(h), and may be used for the management of remediation wastes. However, some stipulations apply, such as:

- CAMUs must be on the same facility
- CAMUs could include areas on the same facility that are not contiguous
- CAMUs are not allowed for operators using interim status
- Could include regulated units in a CAMU
- Placement of remediation wastes in a CAMU does not trigger LDRs, and
- CAMUs apply to remediation wastes, not "as generated wastes."

Some non-RCRA hazardous waste may be subject to management requirements under Subtitle D of RCRA as solid wastes. Subtitle D regulates disposal of solid waste in facilities such as municipal landfills. Therefore, non-RCRA hazardous IDW, such as decontaminated PPE or equipment, may need to be disposed of in a Subtitle D facility depending on state requirements (EPA 1992).

Discharges of aqueous IDW to surface water or POTWs may be required to comply with the CWA federal, state, and local requirements. These requirements may include (EPA 1992):

- water quality criteria
- pretreatment standards
- state water quality standards, and
- NPDES permit conditions.

If IDW contains PCBs, TSCA treatment and/or disposal requirements may apply during its management. TSCA requirements regulate the disposal of material contaminated with PCBs at concentrations of 50 ppm or greater. In addition, TSCA storage requirements may apply that limit the time that PCBs may be stored to one year. If PCB material is mixed with RCRA hazardous waste, it may be regulated by the LDR California list prohibitions (EPA 1992). For further details, see RCRA sections 3004(d)(2)(D) and (E).

If IDW is to be disposed of offsite, which requires the transportation of the material along public roads, Department of Transportation requirements for containerizing, labeling, and transporting hazardous materials and substances may apply (EPA 1992).

State requirements which are legally enforceable and more stringent than federal regulations may be potential ARARs for IDW managed on the site. Substantive State requirements that may be ARARs include state water quality control standards, direct discharge limits, and RCRA requirements promulgated in a state with an authorized RCRA hazardous waste management program (EPA 1992).

Before a waste can be shipped offsite, approval should be obtained for the proposed disposal facility from EPA's Regional Off-Site Policy Coordinator. In addition, for Superfund waste shipments out of state, written notification should be provided to the receiving states (EPA 1992).

OBJECTIVES FOR WASTE MANAGEMENT

The EPA has identified two general objectives that should be considered when managing IDW, which include (EPA 1992):

- waste minimization, and
- managing waste consistent with the final remedy for the site.

Every effort should be made to minimize the generation of IDW to reduce the need for special storage or disposal facilities which provide little or no reduction in site risk relative to the final remediation. IDW can most easily be minimized by selecting investigative methods which generate less waste. For example (EPA 1992):

- To reduce the amount of drill cuttings, boreholes should be drilled no larger than is needed to collect the environmental sample.
- Tools such as cone penetrometers and soil-gas probes, which generate very little waste, can be used to assist site characterization studies.
- Decontaminating reusable equipment is preferred over using disposable equipment.

To avoid the need for separate treatment and/or disposal arrangements, IDW should be considered part of the site and should be managed with other wastes from the site, consistent with the final remedial remedy (EPA 1992).

OFFSITE FINAL REMEDIES

If there is a strong likelihood that the final remedy for a site will involve off-site disposal of the waste, it is recommended that the waste be stored in drums or a covered pile onsite until the final action. Returning the waste to its source is generally less protective, may not be in compliance with ARARs, and would require that the soil be excavated again at the time of remediation. The only time that soil should be returned to its source, when offsite disposal is the most likely final remedy, is if storage at the site is not possible, and this option has been determined to be both protective and in compliance with ARARs (EPA 1992).

The following is an example of how IDW is handled as part of a common offsite disposal remedy:

Example:

"A site involves volatile organic RCRA hazardous wastes that will likely be sent off site for final treatment and disposal. Site conditions are such that temporary storage of IDW is considered protective until the remedial action begins. Because off-site disposal will trigger RCRA disposal requirements such as the LDRs and immediate containerization would be more protective than redepositing into the source area at the time of sampling, the site manager decides to containerize the IDW (and comply with RCRA substantive technical tank and container standards) until the final action is initiated" (EPA 1992).

ONSITE FINAL REMEDIES

If onsite disposal is the most likely final remedy, the EPA expects that sludge or soil IDW will be returned to its source if short-term protectiveness is not an issue. The presumption is that IDW that may pose a risk to human health and the environment in the long term will be addressed by the final action. The IDW can be stored temporarily before redepositing, since storage of RCRA hazardous IDW in containers within the AOC will not trigger the LDRs, as long as the containers are not managed in such a way as to constitute a RCRA storage unit as defined by 40 CFR 260.10. However, the EPA believes that, assuming it is protective, returning sludges and soils to their sources immediately can avoid potential increased costs and requirements associated with storage.

The following are examples of how soils and sludges are commonly handled as part of a common onsite disposal remedy:

Example 1:

"The soil at a site contains wastes that are expected to be stabilized on site during the final remedial action. The site manager determines that sending soil IDW off site is not cost-effective, because off-site disposal would involve testing and transport costs for a relatively small amount of waste. Instead, knowing that the site is secure and that re-disposing the waste at the source will not increase site risk or violate ARARs, the site manager decides to return soil IDW to the source area from which it originated" (EPA 1992).

Example 2:

"A site manager determines that returning highly contaminated PCB wastes to the ground at a site is not protective because of the potential risks associated with the material; instead, the site manager chooses to drum the waste and send it off site (in compliance with TSCA). Off-site disposal may occur immediately or at a later date" (EPA 1992).

Example 3:

"Soil IDW contaminated with a RCRA hazardous waste is generated from a soil boring. The site manager decides to put the IDW back into the borehole immediately after generation, but ensures that site risks will not be increased (e.g., the contaminated soil will not be replaced at a greater depth than where it was originally so that it will not contaminate "clean" areas) and that the contamination will be addressed in the final remedy" (EPA 1992).

Appropriate disposal options for aqueous liquids must be made on a site-specific basis. The parameters that must be considered include (EPA 1992):

- the volume of the liquid
- contaminants present in the soil and groundwater
- whether the groundwater or surface water is a drinking water source, and
- whether the groundwater plume is contained or moving.

If drilling fluids contain significant solid component, they may require special disposal and handling. The following are examples of how aqueous liquids are commonly handled as part of a common onsite disposal remedy:

Example 1:

"A site manager has large volumes of ground water IDW and does not know if it is contaminated. Pouring this IDW on the ground would not be protective, because it may contaminate previously uncontaminated soil or may mobilize contaminants that are present in the soil. Therefore, the site manager stores the water in a mobile tank until a determination is made as to whether the water and soil are contaminated or until the final action" (EPA 1992).

Example 2:

"IDW is generated from the sampling of background, upgradient wells. Because there are no community concerns or evidence of any soil contamination from other sources, the site manager decides to pour this presumably uncontaminated IDW on the ground around the well" (EPA 1992).

Example 3:

"Purge water from a deep aquifer is known to be contaminated with a RCRA hazardous waste. At this site, if this water were poured on the ground, it could contaminate a previously uncontaminated shallow aquifer that is a potential drinking water source and would have to comply with the LDRs. The site manager decides to containerize the water within the AOC and store it until the final remedy" (EPA 1992).

Nonindigenous IDW such as sampling materials, disposable PPE, and decontamination fluids, should either be stored until the final remedy or disposed of immediately. If contaminated, it cannot be disposed of onto the ground because this would add contamination that was not present when activities began at the site. If these types of material are contaminated with RCRA hazardous waste, they must be managed in accordance with RCRA Subtitle C requirements. If PPE is not contaminated, it may be disposed of in an onsite dumpster.

The following are examples of how nonindigenous IDW are commonly handled as part of a common onsite disposal remedy:

Example 1:

"Disposable PPE (e.g., gloves, shoe covers) becomes contaminated with RCRA hazardous waste during the field investigation. The site manager containerizes and disposes of the IDW in compliance with RCRA Subtitle C requirements" (EPA 1992).

Example 2:

"Disposable equipment becomes contaminated during a field investigation. The site manager decontaminates them and sends them to a Subtitle D facility" (EPA 1992).

REFERENCES

Environmental Protection Agency, Guide to Management of Investigation-Derived Wastes, 9345.3–03FS, 1–7, 1992.

Environmental Protection Agency, Management of Investigation-Derived Wastes During Site Inspections, EPA/540/G-91-009, 3–4, 1991.

Environmental Protection Agency, Guidance on Remedial Actions for Superfund Sites With PCB Contamination, EPA/540/G-90/007, 1990.

Appendix:
General Reference Materials

Conversion Table

Measurement Conversions
ENGLISH TO METRIC

Known	Multiplier	Product
LENGTH		
inches	2.54×10^4	micron [=10,000 Angstrom units]
inches	25.4	millimeters
feet	30.48	centimeters
yards	0.9144	meters
miles (statute)	1.6093	kilometers
AREA		
square inches	6.4516	square centimeters
square feet	0.0929	square meters
square yards	0.8361	square meters
square miles (1 square mile = 640 acres)	2.5900	square kilometers
VOLUME		
cubic inches	16.3871	cubic centimeters
cubic feet	0.02832	cubic meters
cubic yards	0.7646	cubic meters
cubic miles	4.1684	cubic kilometers
quarts (U.S. liquid)	0.9463	liters (=1000 cm^3)
gallons (U.S. liquid) (=0.8327 Imperial gal)	3.7854	liters
barrels (petroleum — 1 bbl = 42 gal)	158.9828	liters
acre-feet (=43,560 ft^3 = 3.259×10^5 gal)	1233.5019	cubic meters
MASS		
ounces	28.3495	grams
pounds	0.4536	kilograms
short tons (2000 lb)	0.9072	megagrams = (metric tons)
long tons (2240 lb)	1.0160	megagrams
carats (gems)	0.2000	grams
VOLUME PER UNIT TIME		
cubic feet per second (= 448.83 gal/min)	0.02832	cubic meters per second
cubic feet per second	28.3161	cubic decimeters per second (= liters per second)
cubic feet per minute (= 7.48 gal/min)	0.47195	liters per second
gallons per minute	0.06309	liters per second
barrels per day (petroleum—1 bbl = 42 gal)	0.00184	liters per second

(Dietrich 1982)

Conversion Table (cont.)

METRIC TO ENGLISH

Known	Multiplier	Product
LENGTH		
micron (=10,000 Angstrom units)	3.9370×10^{-5}	inches
millimeters	0.03937	inches
centimeters	0.0328	feet
meters	1.0936	yards
kilometers	0.6214	miles (statute)
AREA		
square centimeters	0.1550	square inches
square meters	10.7639	square feet
square meters	1.1960	square yards
square kilometers	0.3861	square miles (1 square mile = 640 acres)
VOLUME		
cubic centimeters	0.06102	cubic inches
cubic meters	35.3146	cubic feet
cubic meters	1.3079	cubic yards
cubic kilometers	0.2399	cubic miles
liters (=1000 cm^3)	1.0567	quarts (U.S. liquid)
liters	0.2642	gallons (U.S. liquid)
liters	0.006290	barrels (1 bbl = 42 gal)
cubic meters	0.0008107	acre-feet (=43,560 ft^3 = 3.259 $\times 10^5$ gal)
MASS		
grams	5.0000	carats (gems)
grams	0.03527	ounces
kilograms	2.2046	pounds
megagrams (=metric tons)	1.1023	short tons (2000 lb)
megagrams	0.9842	long tons (2240 lb)
VOLUME PER UNIT TIME		
cubic meters per second	35.3107	cubic feet per second (=448.83 gal/min)
cubic decimeters per second (liters per second)	0.03532	cubic feet per second
liters per second	2.1188	cubic feet per minute
liters per second	15.8503	gallons per minute
liters per second	543.478	barrels per day (petroleum —1 bbl = 42 gal)

(Dietrich 1982)

Conversion Table (cont.)

Volume

1 cubic ft	= 7.4805 U.S. gallons	= 6.2321 imperial gallons	= 28.37 liters
1 U.S. gallon	= 0.13368 cubic ft	= 0.83271 imperial gallons	= 3.7825 liters
1 imperial gallon	= 0.16046 cubic ft	= 1.2009 U.S. gallons	= 4.5437 liters
1 liter	= 0.03531 cubic ft	= 0.26417 U.S. gallons	= 0.22009 imperial gallons

1 cubic ft	= 0.028317 cubic meter	= 0.000022957 acre-ft
1 cubic meter	= 35.3145 cubic ft	= 0.00081071 acre-ft
1 acre-ft	= 43,560 cubic ft	= 1,233.5 cubic meters

1 cubic mile = 3.3792 million acre-ft

1 cfs-day = 86,400 cubic ft = 1 cubic ft per second for 24 hr

Volume Conversion Factors

Initial Unit	Coefficient (multiplier) to obtain:					
	Cfs-day	Mil. cu. ft	Mil. gal.	Acre-ft	In. per sq mi.	Mil. cu. meters
Cfs-days	--	0.086400	0.64632	1.9835	0.037190	.0024466
Mil. cu. ft	11.574	--	7.4805	22.957	.43044	.028317
Mil. gal.	1.5472	.13368	--	3.0689	.057542	.0037854
Acre-ft	.50417	.043560	.32585	--	.018750	.0012335
In. per sq. mi.	26.889	2.3232	17.379	53.333	--	.065785
Mil. cu. meters	408.73	35.314	264.17	810.70	15.201	--

Velocity

1 mile per hr	= 1.467 ft per sec
1 mile per hr	= 88 ft per min
1 ft per sec	= 0.682 mile per hr
1 ft per min	= 0.0114 mile per hr
1 ft per sec	= 0.3048 meter per sec
1 meter per sec	= 3.281 ft per sec

Pressure (0° C = 32° F)

1 ft of head, fresh water	= 0.433 lb per sq in, pressure
1 lb per sq in, pressure	= 2.31 ft of head, fresh water
1 meter of head, fresh water	= 1.42 lb per sq in, pressure
1 lb per sq in, pressure	= 0.704 meter of head
1 atmosphere (m.s.l.)	= 33.907 ft of water

Weight

1 cubic ft of fresh water	= 62.4 lb	= 28.3 kg
1 cubic ft of sea water	= 64.1 lb	= 29.1 kg
1 cubic meter of fresh water	= 1000 kg	= 1 metric ton

(Dietrich 1982)

Conversion Table (cont.)

Rates of Flow

1 cubic ft per sec	=	448.83 U.S. gallons per min	=	646,317 U.S. gallons per day	=	.028317 cu meter per sec
1 cubic ft per min	=	7.4805 U.S. gallons per min	=	10,722 U.S. gallons per day	=	.00047195 cu meter per sec
1 U.S. gallon per min	=	0.002228 cubic ft per sec	=	0.13368 cubic ft per min	=	1440 U.S. gallons per day = .000063090 cu meter per sec
1 U.S. gallon per day	=	.000093 cubic ft per min	=	.0006944 U.S. gallons per min		
1 cubic ft per sec	=	1.9835 acre-ft per day	=	723.98 acre-ft per year	=	.014276 cu meter per sec
1 acre-ft per day	=	0.50417 cubic ft per sec	=	365 acre-ft per year		
1 acre-ft per year	=	0.00138 cubic ft per sec	=	0.00274 acre-ft per day		
1 inch per hr on 1 acre	=	1 cubic ft per sec (approx.)				
1 inch per hr on 1 sq mi	=	645.33 cubic ft per sec				

Volume Conversion Factors

Initial Unit	Coefficient (multiplier) to obtain:					
	Cu ft per sec	Gal per min	Mil gal per day	Acre-ft per day	Inches per day per sq mi	Cu meters per sec
Cu ft per sec (cfs)	--	448.83	0.64632	1.9835	0.037190	.028317
Gal per min (gpm)	0.0022280	--	.001440	.0044192	.00008286	.000063090
Mil gal per day (mgd)	1.5472	694.44	--	3.0689	.057542	.043813
Acre-ft per day	.50417	226.29	.32585	--	.01850	.014276
Inches per day per sq mi	26.889	12,069	17.379	53.333	--	.76140
Cu meters per sec	35.314	15,850	22.834	70.045	1.3134	--

(Dietrich 1982)

Unified Soil Classification System

MAJOR DIVISIONS			GROUP SYMBOLS	TYPICAL NAMES
COARSE-GRAINED SOILS More than half of material is larger than no. 200 sieve size.	GRAVELS More than half of coarse fraction is larger than no. 4 sieve size.	Clean gravels	GW	Well-graded gravels, gravel-sand mixtures, little or no fines.
			GP	Poorly graded gravels, gravel-sand mixtures, little or nor lines.
		Gravels with fines	GM	Silty gravel, gravel-sand-silt mixtures.
			GC	Clayey gravels, gravel-sand-clay mixtures.
	SANDS More than half of coarse fraction is smaller than no. 4 sieve size.	Clean sands	SW	Well-graded sands, gravelly sands, little or no fines.
			SP	Poorly graded sands, gravelly sands, little or no fines.
		Sands with fines	SM	Silty sands, sand-silt mixtures.
			SC	Clayey sands, sand-clay mixtures.
FINE-GRAINED SOILS More than half of material is smaller than no. 200 sieve size.	SILTS AND CLAYS	Low liquid limit.	ML	Inorganic silts and very fine sands, rock flour, silty or clayey fine sands, or clayey silts, with slight plasticity.
			CL	Inorganic clays of low to medium plasticity, gravelly clays, sandy clays, silty clays, lean clays.
			OL	Organic silts and organic silty clays of low plasticity.
		High liquid limit.	MH	Inorganic silts, micaceous or diatomaceous fine sandy or silty soils, elastic silts.
			CH	Inorganic clays of high plasticity, fat clays.
			OH	Organic clays or medium to high plasticity, organic silts.
	Highly organic soils		Pt	Peat and other highly organic silts.

NOTES:

1. Boundary Classification: Soils possessing characteristics of two groups are designated by combinations of group symbols For example, GW-GC, well-graded gravel-sand mixture with clay binder.

2. All sieve sizes on this chart are U.S. Standard.

(Dietrich 1982)

Concept and Classification of Soils

DEFINITION: Soil is a natural, historical body with an internal organization reflected in the profile and its horizons, consisting of weathered rock materials and organic matter with the former usually predominant, and formed as a continuum at the land surface largely within the rooting zones of plants.

HYPOTHETICAL SOIL PROFILE:
with notations for master horizons

pre-1980 nomenclature	current nomenclature	
01	Oi	Loose leaves and organic debris, largely undecomposed.
02	Oe	Organic debris, partially decomposed.
A1	Ah	A dark-colored horizon of mixed mineral and organic matter and with much biological activity.
A2	E	A light-colored horizon of maximum eluviation; prominent in some soils but absent in others.
A3	EB	Transitional to B but more like A (or E) than B; may be absent.
B1	BE	Transitional to A (or E) but more like B than A (or E); may be absent.
B2	B	Maximum accumulation of silicate clay minerals or of sesquioxides and organic matter; maximum expression of blocky or prismatic structure; or both.
B3	BC	Transitional to C but more like B than C; may be absent.
C	C	Weathered parent material; occasionally absent; formation of horizons may follow weathering so closely that the A or B horizon rests on consolidated rock.
R	R	Layer of consolidated rock beneath the soil.

(Dietrich 1982)

Radiological Decay Chains

Uranium-235

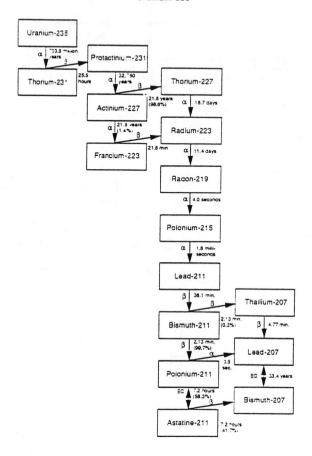

NOTES:
The times shown are half-lives.
The symbols α or β indicates alpha and beta decay.

Radiological Decay Chains
(continued)

Uranium-238

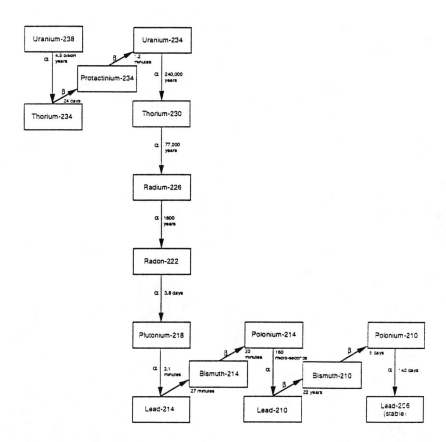

NOTES:
Only the dominant decay mode is shown.
The times shown are half-lives.
The symbols α or β indicate alpha and beta decay.
An asterisk indicates that the isotope is also a gamma emitter.

Radiological Decay Chains
(continued)

Thorium-232

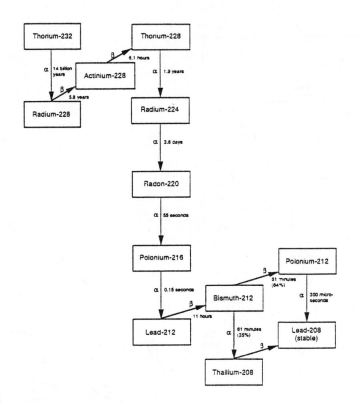

NOTES:
The times shown are half-lives.
The symbols α or β indicate alpha and beta decay.
An asterisk indicates that the isotope is also a gamma emitter.

REFERENCES

Dietrich, R.V., Dutro, J.T., Jr., and Foose, R.M., AGI Data Sheets, 1982.

Index